WITHDRAWN

IN SEARCH OF THE
AFRICAN WILD DOG

The right to survive

In search of the African Wild Dog

The right to survive

ROGER & PAT DE LA HARPE

SUNBIRD
PUBLISHERS

DEDICATION

*To Marc and Rose, Vaughan and Liz, Willem and Sandy –
with love and thanks.*

Sunbird Publishers (Pty) Ltd
An imprint of Jonathan Ball Publishers
PO Box 6836, Roggebaai, 8012, Cape Town, South Africa

www.sunbirdpublishers.co.za
Registration number: 1984/003543/07

First published in 2009
Copyright © text Pat de la Harpe
Copyright © photography Roger de la Harpe
Copyright © illustration/cartography Sunbird Publishers
Copyright © published edition Sunbird Publishers

PUBLISHER Ceri Prenter
EDITOR David Bristow
DESIGN Pete Bosman
ILLUSTRATIONS Adam Carnegie, André de la Rosa
PROOFREADER Kathleen Sutton
REPRODUCTION BY Resolution Colour Pty Ltd
PRINTED AND BOUND BY Tien Wah Press (Pte) Ltd, Singapore

Standard edition ISBN 978-1-91993-811-0
Collector's edition ISBN 978-1-92028-912-6
Sponsor's edition ISBN 978-1-92028-913-3

All rights reserved. No part of this publication may be reproduced, stored in a retrieval system or transmitted, in any form or by any means, electronic, mechanical, photocopying, recording or otherwise, without the prior written permission of the copyright owner(s).

Historical photos property of the authors, with the exception of King Lobengula, page 106, and President Paul Kruger, page 136, courtesy of MuseumAfrika.

Title page The wild dog, *Lycaon pictus*, is Africa's second most endangered carnivore, after the Ethiopian or Simian wolf *Canis simensis*.

This spread Wild dogs love water (even though they can live without it) and as long as it is not deep will enjoy playing in it.

Overleaf The Sabi Sand Game Reserve forms part of the greater Kruger National Park, the most significant range for wild dogs in South Africa.

Contents

SPONSOR'S FOREWORD 9
PREFACE 10
ACKNOWLEDGEMENTS 12
DISTRIBUTION MAP 13

ONE
The painted wolf 14
(Lycaon pictus)
SOUTH AFRICA

TWO
Makanyane 42
(Tswana)
NORTH WEST BUSHVELD

THREE
Amankentshane 72
(Zulu)
ZULULAND

FOUR
Matlhalerwa 102
(Venda/Tswana)
LIMPOPO VALLEY

FIVE
Mahlolwa 132
(Shangaan)
GREATER KRUGER NATIONAL PARK

REFERENCES 158
INDEX 159

Sponsor's Foreword

In 1994 Sasol was approached by De Wildt's Ann Van Dyk to become involved in their wild dog research, breeding and release programme. Already a supporter of education about and conservation of birds, Sasol decided to take on the cause of the endangered African wild dog. The De Wildt Cheetah and Wildlife Trust has taken on the responsibility of conserving these magnificent predators outside of conservation areas, as the conservation authorities in South Africa do not have the capacity or the funding to do so.

With Sasol's support, the Trust launched a project to focus on the conservation and survival of the species at least within South Africa.

Our funding of a wild dog camp, quarantine facility and vaccination research has largely enabled De Wildt, in collaboration with many other institutions, to realise their conservation objectives. We are very pleased with the role that tours through the Sasol wild dog camp have played in educating thousands of South Africans and foreign visitors to the plight of this species. With estimates of fewer than 3 000 dogs in southern Africa, the race for survival of wild dogs may hinge in part on dispelling negative perceptions about this species.

We are also grateful that our funding has been able to positively impact on the research, breeding and release programme, and would like to thank all those organisations which have been involved. Sasol is very pleased that ecotourism and occupancy in South Africa's national parks have been positively influenced through the wild dog release programme. We hope that our support of this publication, *In Search of the African Wild Dog*, which tells the conservation story thus far, will further contribute towards ensuring the survival of the 'painted wolf' – the wild dog.

PAT DAVIES
CHIEF EXECUTIVE
SASOL

Opposite and above Playing, resting or hunting, wild dogs interact continually with other members of their pack.

Preface

Mahlolwa: the Shangaan word for an outcast. For an animal to be regarded as an outcast – for it to be hounded mercilessly or shunned – it must have committed some dreadful crime.

Too often we hear that wild dogs, also known as 'painted wolves', are the most callous and bloodthirsty of all carnivores. That they are fast and efficient killers, taking out the weak and sick – a necessary factor in balancing the wildlife populations – is often overlooked. Sadly, we have ignored these facts and brought the status of the wild dog to an endangered species with a possible future of complete extermination.

There are visionaries who work tirelessly to conserve threatened wildlife species and reverse this unacceptable trend. Sasol, for one, has been a farsighted forerunner in wild dog conservation and has played a major role in the development of the De Wildt Cheetah and Wildlife Centre wild dog programme.

The partnership began in 1994 with the building of the Sasol camp and Sasol den at De Wildt. The camp enabled us to study the social behaviour of the dogs as a pack, with the view of releasing captive-born dogs back into the wild, while the den allowed us to observe and film the birth of the puppies.

This spread The fluid movement of a wild dog as it lopes across the veld is incredible to see and reflects its amazing athleticism.

This study has enabled me to fulfil a dream of returning captive-born wild dogs to their natural habitat. It was accomplished by combining captive with wild-caught dogs prior to release, although we were constantly advised that to do so would be plain crazy! But thanks to Sasol, backed by its then executive director Jan Fourie, and a group of dedicated conservationists, a pack of three wild-caught males and three captive-born females was released in the Madikwe Game Reserve. This was a great success, and today the reserve is known for its thriving wild dog population.

So we persevere with the bonding and release of wild and captive-born dogs and are pleased with the successes in a number of other reserves. However, I do foresee that, in the long run, free-roaming wild dogs on unprotected land possibly will not survive in South Africa. Our only recourse is to maintain strong, genetically viable breeding lines in captive populations from which to replenish the free-roaming metapopulation as it becomes depleted.

Through our work at De Wildt, I am humbled and proud to be able to contribute to the preservation of the wild dog and I am sure that this book will highlight the cause of a truly misunderstood species. Readers will discover – as I have – the joy and thrill of knowing the painted wolf.

ANN VAN DYK
FOUNDER AND DIRECTOR
DE WILDT CHEETAH CENTRE

Acknowledgements

Our sincere appreciation to Sasol Ltd for their interest in and sponsorship of our wild dog book, with special mention of Richard Hughes, Sponsorship Manager, and Jan Fourie, now retired, who first approached us to tackle this project several years ago and who we hold responsible for our 'wild dog immersion' over the past year!

Our thanks to the management and staff of the various lodges in the Madikwe Collection whose assistance and hospitality were invaluable. We would particularly like to mention the following people: Kate Naughton and her marketing team, Carmen van den Berg and Patience Bugatsu of Thakadu River Camp, Heidi Janson and Gavin Tonkinson and the team at Tuningi Safari Lodge and Moremi Keabetswe of Buffalo Ridge Safari Lodge.

Our thanks also to Peter Leitner of North West Parks and Tourism Board and to ecologists Declan Hofmeyr and Steve Dell for their unstinting help and advice.

A big thanks to Ann van Dyk, Vanessa Bezuidenhout and Prof. Henk Bertschinger of De Wildt Cheetah and Wildlife Trust, Lee Dicks and Adrian Bantich of Tintswalo Safari Lodge in the Manyeleti, Brett Gehren of Rhino Walking Safaris in the Kruger National Park and Lente Rhoode of the Hoedspruit Endangered Species Centre. In the Northern Tuli Game Reserve our appreciation goes to Nick Hiltermann and Jacques van der Merwe of Tuli Safari Lodge, David Evans and Pete and Jane le Roux of Mashatu, Shane Pinchen of Nitani and to Craig Jackson for his tireless assistance in the field. Thanks too to Ada Hardbattle of Buitsivango farm near Ghanzi in Botswana for her information on Bushman mythology.

Our appreciation also to Brendan Whittington-Jones, Carla Graaff, Mariana Venter and Zama Zwane of the Endangered Wildlife Trust, Dave Robinson and Paul Havemann of Ezemvelo KZN Wildlife and Simon Pillinger of Thanda Private Game Reserve. To Dr Gus Mills and Dr Markus Hofmeyr of South African National Parks, Harriet Davies-Mostert of the Wild Dog Advisory Group, Warwick Davies-Mostert of Venetia Limpopo Nature Reserve, Michael Somers for his pile of reference material and to Frans Prins for his information on Bushman rock art, a big thanks.

Last but not least to the Sunbird Publishing team – Ceri Prenter, Peter Bosman and David Bristow – it was a pleasure working with you.

HOWICK, SEPTEMBER 2009.

Where the Wild Dogs Are

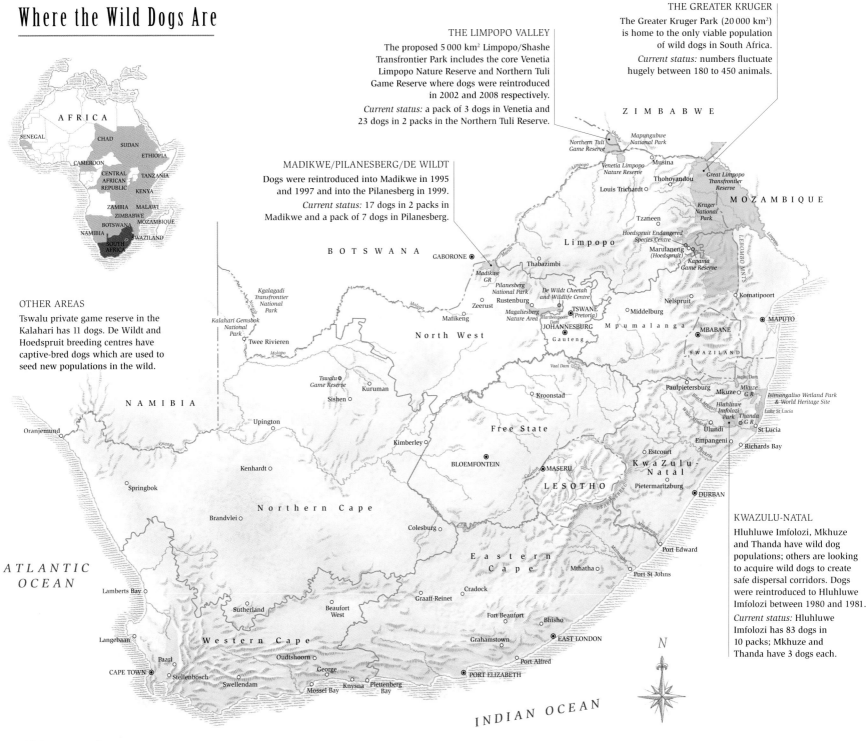

THE LIMPOPO VALLEY
The proposed 5 000 km² Limpopo/Shashe Transfrontier Park includes the core Venetia Limpopo Nature Reserve and Northern Tuli Game Reserve where dogs were reintroduced in 2002 and 2008 respectively.
Current status: a pack of 3 dogs in Venetia and 23 dogs in 2 packs in the Northern Tuli Reserve.

THE GREATER KRUGER
The Greater Kruger Park (20 000 km²) is home to the only viable population of wild dogs in South Africa.
Current status: numbers fluctuate hugely between 180 to 450 animals.

MADIKWE/PILANESBERG/DE WILDT
Dogs were reintroduced into Madikwe in 1995 and 1997 and into the Pilanesberg in 1999.
Current status: 17 dogs in 2 packs in Madikwe and a pack of 7 dogs in Pilanesberg.

OTHER AREAS
Tswalu private game reserve in the Kalahari has 11 dogs. De Wildt and Hoedspruit breeding centres have captive-bred dogs which are used to seed new populations in the wild.

KWAZULU-NATAL
Hluhluwe Imfolozi, Mkhuze and Thanda have wild dog populations; others are looking to acquire wild dogs to create safe dispersal corridors. Dogs were reintroduced to Hluhluwe Imfolozi between 1980 and 1981.
Current status: Hluhluwe Imfolozi has 83 dogs in 10 packs; Mkhuze and Thanda have 3 dogs each.

African overview

Previously found throughout most of the continent, the wild dog has disappeared from much of its former range and is practically non-existent in west, central and northeast Africa. A few remnant populations in southern Africa and the southern part of east Africa remain. Their current IUCN Status is listed as 'endangered with a decreasing population trend'. It is estimated there are only between 3 000 and 5 500 wild dogs left in the whole of Africa. South Africa, one of a few African countries with a viable wild dog population, has a mere 500 dogs.

These dismal statistics are due largely to their increasing contact with humans. Rapid urbanisation throughout the continent has led to growing encroachment into the wild dogs' natural ranges, which are being sliced up and isolated by the human footprint. Contact with humans means contact with domestic animals and diseases like canine distemper and rabies, which have been known to wipe out entire packs. Perhaps their greatest threat, however, continues to be the belief that they are wanton killers and are still frequently shot or poisoned whenever they are spotted.

ONE

The painted wolf

(Lycaon pictus)

SOUTH AFRICA

Left Wild dogs run effortlessly. With their slender frame and long thin legs, they are built for speed and can reach an estimated 60 kilometres an hour during a hunt. They use both pace and stamina, rotating lead, to exhaust their prey.

Our wild dog book first stirred into life in 2000 when Jan Fourie, an executive director of Sasol at the time, approached us to write about 'their dogs'. After some discussion we established he was referring to the two wild dog packs that had been released into the Madikwe Game Reserve in 1995 and 1997 respectively, with the help of a sponsorship from his organisation. On both occasions the packs had been unique in that they were made up of a combination of captive-bred and wild-caught individuals – something that had not been attempted before and which proved to be a great success. Furthermore, the captive-bred dogs had come from the De Wildt Cheetah and Wildlife Centre in the Magaliesberg. That organisation has a long association with Sasol and has, over the years, played an educational role in helping to dispel the generally negative perceptions that people have of the African wild dog – Lycaon pictus.

Once we had started our research, however, it became obvious that to tell the De Wildt/Madikwe story it would be necessary to extend it to include other conservation areas in South Africa and adjoining areas that subscribe to the concept of a wild dog 'metapopulation'. This involves the translocation of dogs between different reserves to simulate their natural dispersal and prevent inbreeding. The Kruger National Park, home to the only genetically viable group of wild dogs left in South Africa, is essential to the story. Equally important is the vast area of private game reserves on the park's unfenced western boundary, which often plays host to some of the Kruger packs. And so it was that we embarked on a journey, which was to become as emotional as it was physical, that criss-crossed the country from the arid northern regions of Limpopo province to the sub-tropics of KwaZulu-Natal, in an ever-widening search for these most elusive of predators.

Opposite As unique as a human fingerprint, every wild dog has its own distinctive colour pattern. Their beautiful markings come in combinations of black, white and tan, although almost without exception they have white-tipped tails. While both males and females look alike, the male is generally larger.

A Determined Survivor

The African wild dog, *Lycaon pictus*, can trace its ancestry back millions of years. It is one of the Canidae family, the members of which include wolves, coyotes, foxes and dingoes and occur in a variety of habitats across the world. Of the five species in southern Africa – the wild dog, black-backed jackal, side-striped jackal, Cape Fox and bat-eared fox – the wild dog is the largest and is purely meat eating. While there is a general similarity between the various canids worldwide, the African wild dog differs from the others in a fundamental way: it belongs to the genus, *Lycaon*, which formed a new branch on the family tree some three million years back and subsequently evolved completely independently. Today it is the only survivor of this unique line and, because of its genetic difference, is unable to interbreed with any of its canid relatives or even with the domestic dog. The species was originally named *Hyaena picta* by Temminck (first director of the National Natural History Museum in Leiden, Holland), in 1820, from a specimen from coastal Mozambique.

The derivation of the wild dog's scientific name, *Lycaon pictus*, is as intriguing as the animal itself. The word 'lycaon' comes from an ancient Greek myth, in which the first King of Arcadia offered Zeus (king of the Gods) a plate of human flesh, to test his omnipotence. Zeus pushed the plate away in disgust and duly turned the king and his 50 impious sons into wolves. This myth is intertwined with another that tells of an ancient cannibalistic sect in pre-Hellenistic Greece, from which the legend of the werewolf (lycanthropy) originates. *Pictus* means painted in Latin and refers to the painted effect of the animal's coat – together these words, lycaon and pictus, make up the evocative scientific name of the African painted wolf.

To further add to their mystique they are known locally as the Cape hunting dog, *wildehond* (Afrikaans), *!//haru* (Bushman), *makanyane* (Sotho/Tswana), *matlhalerwa* (Tswana), *amankentshane* (Zulu) and *mahlolwa* (Shangaan), to mention a few. All these names describe the same animal – a sleek and slender predator, weighing up to 26–28 kg when fully grown, with long thin legs, a large head, rounded black ears, a white-tipped tail and a marbled coat in various combinations of black, white and tan.

Each dog has its own distinctive colour pattern: its designer pelt provides excellent camouflage on the grassy plains, in the open woodlands and bushy savannas it frequents. Their unique markings have proved invaluable to researchers as a way of identifying individuals in study groups. Interestingly enough, the wild dogs of southern Africa are generally larger and lighter in colour than their cousins further north (average 18 kg). This is believed to be connnected to the fact that the principal prey in east Africa is the smaller Thomson's gazelle (20–27 kg), while in southern Africa it is the impala (45–65 kg).

Perhaps the most successful hunter in Africa, it is ironically also one of the most endangered. In the past wild dogs were found in diverse habitats across the continent, with the exception of densely forested and extremely arid areas. There is a report in the *East African Wildlife Journal* dated 1970 that a climber – none other than the famous explorer Wilfred Thesiger – encountered several wild dogs on the slopes of Kilimanjaro, at 5 894 metres above sea level. Certainly in recent times both lion and hyena spoor have been seen above 4 000 metres.

Tragically, today they have disappeared from much of their former range and are practically non-existent in west, central and northeast Africa, with a few remnant populations in southern Africa and the southern part of east Africa. Even in these areas their continued presence is under severe threat and their current IUCN status is listed as 'endangered with a decreasing population trend'. It is estimated that there are only between 3 000 and 5 500 wild dogs left in the whole of Africa. South Africa is one of a few African countries with a viable wild dog population and can boast a mere 500 of these animals!

These dismal statistics are due largely to their increasing contact with humans. Rapid urbanisation throughout the continent has led to growing human encroachment into the wild dogs' natural ranges, placing huge pressures on their survival: their principal requirement is that they need large areas in which to roam, hunt and form new packs. Wild areas are increasingly sliced up and isolated by the human footprint. Additional roads, fields, fences and alteration of the natural habitat naturally follow

THE PAINTED WOLF

Above Although there are no recorded wild dog attacks on humans, it is quite daunting to be caught unawares in the bush by a pack of inquisitive dogs, as we were in this instance in the Northern Tuli Game Reserve.

human settlement. Increasing numbers of wild dogs are being knocked down and killed by fast-moving vehicles. Contact with humans also means contact with their domestic animals and in turn to diseases such as canine distemper, rabies and anthrax, which have been known to wipe out entire packs of wild dogs at a time. Perhaps their greatest threat, however, continues to be people's belief that they are wanton killers, no doubt based on the dogs' incredibly high success rate while hunting – they hunt and kill successfully almost every day – as well as their technique of disembowelling their prey on the run. Reviled throughout Africa for many years, they are still frequently shot or poisoned whenever they are spotted.

The vilification of wild dogs is not a recent phenomenon. The writings of early pioneers and hunters in the second half of the 19th century and the early part of the 20th century are littered with misconceptions and denigrating comments about the dogs. Consider the following:

> And when the game is very thin in a district they know the dogs have been through, for these terrible brutes hunt in packs ... (A Scrobie, *Animal Heaven*, 1953).

> It was a dark-coloured, wolf-like beast, with a bushy tail tipped with white, and roamed the country in fierce and voracious packs, working wholesale destruction among sheep and goats and calves, as it passed. (AT Bryant, *The Zulu People*, 1949).

> Of all the killers in Africa, the wild dog is the most cruel and relentless, one that never gives up the chase until his victim lies dead before him. (C Napier, *Killers and Big Game*, 1966).

Many of these early hunters mention that when on horseback they were often followed by wild dogs for long distances and, while it was extremely unnerving, they were never attacked.

Above There is a sense of suppressed energy in a wild dog, even when it is standing still. It takes very little for them to dart off and leave you staring at an empty space when, seconds before, they filled your view finder.

Great white exterminators

Wild dogs did not escape the wholesale killing of Africa's game that occurred throughout the continent during this period of colonial exploration and settlement. The idea of the 'great white hunter' was romanticised in the British and European press, the perception of fearless men striding through savage lands turned many into legends and urged others to follow suit. (Indeed names like Frederick Selous, William Baldwin, George 'Elephant' Phillips and William Cornwallis Harris are still familiar to us.) The seemingly endless herds attracted the attention of not only the so-called ethical hunter but also the unscrupulous; those who greedily sought to line their pockets from the proceeds of the indiscriminate slaughter of wild animals. Adventurer Cecil Napier mentions wild dogs in this context in his book *Killers and Big Game*:

> There were several professional crocodile hunters camped near a river, who made a meagre living out of the soft bellies of the saurian's skins... These men made a little extra money shooting wild dogs and as they had to hunt in order to keep the camp supplied with fresh meat, it paid for an otherwise wasted day.

> ... It is a strange thing that wild dogs always come back to sniff at the body of a slain dog provided that no human has been seen by them. The sound of rifle fire does not seem to alarm them as it does other animals, and if the hunter is experienced and knows how to conceal himself and keep absolutely motionless, it is possible to exterminate fifteen or twenty dogs before they finally decide to retreat beyond range.

> Not all professional hunters are "sportsmen" in the accepted sense of the word. Some of them would deliberately wound a big buck and leave it to die in the bush or be killed by a predator such as a lion or other beast, following up the spoor of the wounded buck the next day in the hope of finding lions feeding on it, for a lion's skin was worth five pounds and even more for a good leopard. Wild dogs might also be there, or hyenas, both of which carried Government bounties.

Above Impala are on the menu of most large predators. In Kruger these antelope make up around 90 per cent of wild dogs' prey. Spotted hyenas are a constant irritation as they regularly muscle in to steal the dogs' kills.

The Devil's dog

Unfortunately for Africa's game there was yet another sort of 'hunter' who emerged around this time. These were privileged young men, bored with their lifestyle at home in Europe and England and intent on proving their manhood by killing as many wild animals as possible, just to show they could. Perhaps the worst offenders were British officers doing military duty in India, who came out to Africa to hunt during their long leave. The blood lust of some of these hunters knew no end and they were known to stop shooting only when their horses dropped from exhaustion or they ran out of ammunition – one such individual bragged he had a barrel of gunpowder on his wagon and 18 000 bullets! As the plains rang with the sound of gunfire, the carcasses of thousands of antelope, zebras and elephants were left to rot under the hot African sun, along with the corpses of the many predators that followed the slaughter – lions, leopards, cheetahs, hyenas and wild dogs.

But for the wild dog this was not the end of their persecution. For many years they were classified as vermin by provincial administrations in South Africa and rewards were offered for their elimination, which varied from 5s for a tail to 7s 6d for the whole animal. Bounties for leopards were 10s and for jackal and 'lynx' (caracal) 5s. Between 1911 and 1934 more than £369,000 was spent on destroying vermin across the country.

As if this was not enough, wild dogs were systematically eradicated from national parks and other protected areas by wildlife managers, the general consensus at the time being that in order to build up the region's dwindling herds of medium-sized antelope (thought of as 'royal game' by that era's generation of British-schooled rangers) it would be necessary to get rid of those predators that showed a preference for impala, kudu, reed buck and the like. Colonel James Stevenson-Hamilton, first warden of the Sabi Game Reserve (later to become the Kruger National Park), reported in his *Wild Life in South Africa*:

> In spite of rigorous control, wild dogs remained fairly numerous up to about 1927, when at first slowly, and then more rapidly, their numbers lessened. In 1931 the shooting of the species was discontinued by the staff...

Harry Wolhuter, legendary game ranger and Stevenson-Hamilton's right-hand man, acknowledged his role in shooting these so-called vermin in the early 1900s.

That the wild dog survived any of this is nothing short of miraculous, but the species today is in dire straits. In South Africa a handful of isolated groups remain, scattered across a few protected areas.

To add to their tenuous situation, a new and ugly phenomenon has raised its head – over the past five years a growing, unregulated trade in wild dogs with the East is further depleting their numbers by some 100 dogs a year, which could impact negatively on the species. There is no conservation value in this trade as it is purely for commercial purposes, the dogs by all accounts being caged in awful conditions to be gawked at by passing tourists.

Top Colonel James Stevenson-Hamilton was appointed first warden of the Sabi Game Reserve/Kruger National Park in 1902. At first not sure of what his role was to be he was told by his superior, Sir Godfrey Lagden, to 'go down there and make yourself thoroughly disagreeable to everyone.'

Right Professional hunter Frederick Selous (centre) epitomised the Great White Hunter and became the darling of the British press.

THE PAINTED WOLF

Above With its tail low and its ears back, this dog assumes a characteristic stalking pose as it emerges from long grass in the Madikwe Game Reserve. While they adopt this position when first approaching potential prey, it is also often used during play activity between pack members, usually prior to one dog sneaking up and pouncing on another.

Above Harriet Davies-Mostert and her team at the Endangered Wildlife Trust have numerous programmes in place to inform people about the endangered predicament of wild dogs and so try to increase their chances of long-term survival. These fascinating predators are a huge drawcard for tourists wherever they occur in southern Africa.

Opposite Prior to going out to hunt, a rallying call of twitters and squeals spreads between pack members, the rising intensity rousing any that have been dozing or languidly lolling about. Wild dogs usually hunt in the early morning and late afternoon.

Fortunately for this remarkable animal, in southern Africa several influential organisations which are committed to conserving *Lycaon pictus*, have rallied to the cause.

A co-ordinated management plan has been put into place and a working group called The Wild Dog Advisory Group (with the acronym WAG) has been formed. The main purpose of the group is to establish several small populations of wild dogs in different locations and to manage them as a 'metapopulation'. In other words, to artificially link the different groups in different areas of the country by translocating individuals from one population to another in order to try to imitate the natural movement of dogs. This effects the transfer of genes between packs to prevent inbreeding and promote healthy animals.

This project has had several successes over the past few years. Current chairperson of WAG, Harriet Davies-Mostert, comments:

Since 1997, through a series of reintroductions, wild dog populations have been established in several reserves around the country from as far afield as northern KwaZulu-Natal to northern Limpopo province. These small populations are managed as a single population, by moving wild dogs between areas. Between 1998 and 2006, 66 founder animals were used to establish nine such sub-populations, and the metapopulation reached a peak of 264 animals in 17 packs in June 2005.

Although the process of establishing sub-populations has been challenging, the existence of an advisory panel to guide decisions has greatly facilitated the success of the metapopulation strategy. 'The experiences we have gained over the past 10 years have improved our technical capacity to capture and translocate wild dogs, and this is likely to inform wild dog conservation management elsewhere in Africa,' says Davies-Mostert.

A big issue, however, remains the need for more land – larger areas where packs can roam freely without being managed as intensely as they are at present. The proposed transfrontier parks between South Africa and its neighbouring states (some of which are already a reality) will go a long way to providing bigger sanctuaries for wildlife. If this is the case then things might for once go in favour of the wild dogs. It is true their lot has already improved to some extent.

The Power of the Pack

Watching a pack of wild dogs loll intertwined as they rest up in the heat of the day, co-ordinate as a team at a hunt or rush back as one to the den to feed their waiting pups, gives us pause to think. There is nothing really that beats teamwork in a world that has little time for sentiment.

Africa's most efficient hunter is also its most sociable. Wild dogs live together in groups of about six to 30 members (with an average number of 10) and while individually they are rather insignificant among the large predators, their strength and presence lies in the tight social structure of a pack. Close interaction between members of the pack ensures their success as a team and boosts their ability to nurture their young while still hunting effectively.

A pack is usually made up of a dominant pair, the alpha male and female, subordinate adults and various offspring from past and present litters. In a pack all the females are related to one other, as are all the males, but there are no blood ties across the

sexes. As a rule, only the alpha male and female breed and the remaining dogs provide support for the couple and their young. While the dominance of the two top dogs usually inhibits reproduction in other pack members, DNA tests have revealed that on occasion the alpha male is not always the father of all the pups in a litter and that they can be the progeny of several males in the group. From time to time a second female also breeds and the survival of her young depends not only on the amount of food the dogs are able to bring back to the den, but also on the behaviour

Opposite There are strong social bonds between members of a pack, every dog instinctively responsible for the other. This is most evident when the pack has denned and the pups emerge from time to time during the day. The adults rush forward in a frenzy of delight, intent on having some contact with them.

Below It is usually only the alpha female that breeds, but occasionally a second female will also produce a litter. Suckling takes place close to the den. The mother has six to eight pairs of teats, the largest number of any African predator.

of the alpha female, who may either kill the other's offspring, adopt them as her own, or share in the mothering of them.

New packs are formed when, at about two and a half years old, a group of dogs of the same sex breaks away from one pack and goes in search of a splinter group of the opposite sex from another pack. This gives subordinate dogs that were unable to breed in their natal pack a chance to produce offspring in their new one – provided they are lucky enough to become the alpha pair.

Wild dogs have an innate sense of responsibility towards other pack members. They will look after the dominant female during her confinement, guard the den, mind the pups and bring back food to feed the mother, pups and babysitters after a successful hunt. Feeding involves the regurgitation of meat from the kill and this response is stimulated by the pups, which lick and prod the muzzles of the returning dogs as they beg for food, amidst intense and excited squeals and twitters. The offspring are always ravenous and the adults need to work hard to keep them fed. Indeed, care for the youngsters continues even after they are old enough

This spread The den is usually located in an old aardvark hole and some packs return to the same area or even the same den year after year. An average litter usually numbers about 10 and the adults have to work hard to keep them fed. After the reintroduction of wild dogs to the Northern Tuli Game Reserve in April 2008, these pups, born in a thicket next to the Limpopo River, were the first litter in the area after decades of local extinction.

This page Wild dogs hunt as a group: once they have run down their prey, the leading dogs lift their tails into the air, the white tip indicating to the rest that a kill has been made. As they tear at the carcass they make an intense twittering sound. This pack of dogs in Pilanesberg game reserve devoured a mature male impala in a matter of minutes, leaving only the horns, skin and large bones behind.

to join in the hunt, when the adults, themselves hungry, will stand back from a carcass and allow them to first eat their fill.

In southern Africa breeding takes place in midwinter. After a gestation period of about 70 days, a litter of generally around 10 pups is born, but this number can be as many as 19. The den is usually located in an old aardvark hole and some packs have favourite denning areas, returning to the same neighbourhood or even to the same den year after year.

The survival rate of the pups is not high and, in spite of the adults' extraordinary care, they fall prey to other predators, disease or even starvation if the dogs are unable to bring sufficient meat back to the den. It is estimated that only 50 per cent of pups live longer than one year. Large packs have a better success rate of rearing youngsters, as they are generally better able to supply more food. Life is one long game for the pups, their vigorous play a training ground for their future as hunters and their antics are indulgently tolerated by the adults. After about two and a half months they are weaned and then, when they are able to run with the pack a few weeks later, the den is abandoned and the dogs resume a nomadic lifestyle.

Wild dogs have huge home ranges, which can cover areas of between 500 and 2 000 square kilometres. More often than not the ranges of different packs overlap and in the event of two packs bumping into each other, the larger group usually chases off the smaller. The size of their range normally depends on the presence of game and when prey is scarce the dogs travel extensively in their search for food. They are superb hunters and operate as a team during the hunt, relying on their speed and stamina to exhaust their prey. They pull down small to medium-sized antelopes with ease and are able to kill animals much bigger than themselves, including kudu and waterbuck.

In southern Africa they prey largely on impala and in east Africa on Thompson's gazelles, mainly because these are the most abundant species in those areas. In east Africa, however, the dogs appear to hunt wildebeest and zebra more often than those further south, which requires considerable skill as these animals are heavy and powerful and test the strength of much larger predators such as lions and hyenas. Zebras can be especially aggressive and defend themselves with vicious kicks, their flaying hooves potentially lethal.

On average, wild dogs have nearly an 80 per cent success rate during a hunt, which is a remarkable statistic and far outstrips that of a pride of lions, which are successful only 30 per cent of the time. They hunt in the early morning and again in the late afternoon, as well as on moonlit nights when they can see potential prey – they generally use sight rather than sound or smell to locate their quarry. While wild dogs are known to change strategy according to the terrain and the species they are chasing, the leader usually selects a victim, focuses on it and is joined by the rest of the pack that follows the same animal, in spite of the scattering herd around them. Large packs sometimes split up during a hunt and bring down several antelope at a time. The dogs are adept at spotting an animal that is out of condition, old or sick and therefore low on energy and easier to chase.

It is probably their method of killing that has given the wild dogs their reputation as ruthless and indiscriminate killers. Once they have exhausted their quarry they disembowel it, tearing at its vital organs while it is still standing, which is not pretty to see. However, death comes quickly and the animal dies within seconds, unlike the strangulation method used by Africa's big cats – lions, leopards and cheetahs – that takes much longer to dispatch the prey. Such is the co-operation between pack members that no aggression occurs around the carcass. Any problem arising out of the competition for meat is averted as potential rivals become submissive, a tactic peculiar to canids and to *Lycaon* in particular.

Overleaf left Displays of submission, begging and nuzzling are learnt by pups at the den and are carried into adult life. This behaviour is an essential part of a pack's cohesion and is used before a hunt or after its members have become separated. Competition for food also involves begging rather than open aggression.

Overleaf right Large packs have the best chance of survival as there are more dogs available to hunt and bring food back to the den. It also means an adult can stay behind to watch over the pups, as well as any wounded or sick dog that could slow down a hunt and jeopardise chances of a kill.

In search of the African Wild Dog

The Painted Wolf

Those darn cats

When members of a pack become active before a hunt, or when they meet up after they have been separated, they go through a greeting ceremony that involves displays of begging, cringing and nuzzling, accompanied by an excited twittering that is almost birdlike in sound. Apart from the latter they have a range of calls that includes whining and squealing, an owl-like 'hoo' as a contact call and a short, deep bark or growl when they are alarmed. This continual re-affirmation of the group hierarchy is critical for the efficient operation of a pack. Submission and ritualised begging, learnt as pups at the den to get the adults to regurgitate food, is carried into later life and is an essential part of wild dog society.

The pack is able to defend itself against most predators but it will as far as possible avoid any contact with lions, as a confrontation between them almost certainly ends in severe injury or death to several of the dogs. While the enmity between these two species is not as legendary as that between lions and hyenas, wild dogs appear to actively avoid areas where lion numbers are high. (This avoidance also extends to the location of their den.) Given half a chance, a pack of dogs will tree a leopard and will keep it trapped amidst the branches for hours. On the other hand cheetahs pose little threat to them, likewise smaller carnivores such as serval, jackal and caracal.

Spotted hyenas, however, are a constant problem. These notorious scavengers regularly try to steal wild dog kills; while the dogs can chase off single hyenas with ease, when faced with large numbers they are invariably forced to surrender their meal. Not everything goes the hyenas' way in these clashes though and in spite of their much larger size, they become stressed by the dogs' co-ordinated mobbing as, unlike their smaller opponents, every hyena looks out only for itself.

Wild dogs are often mistaken for spotted hyenas and at a cursory glance there is a vague similarity in the colour and patterns of their coats and the shape of their bodies. Both animals inhabit the open plains of the savanna, but the hyena can attain a mass three times that of a wild dog. Perhaps the most significant difference though is that the wild dog, unlike the hyena, scavenges only very occasionally and prefers fresh meat to carrion. The early pioneers spoke of 'hyena dogs' and it was thought at one time that the wild dog was the connecting link between the wolf of Europe and the hyena of Africa.

The dogs' instinct for danger extends to crocodiles and even their love of water will not overcome this fear of them. Their caution as they approach any water that harbours crocodiles is particularly evident among packs that occur in the Okavango Delta in Botswana. Stevenson-Hamilton mentions the behaviour of wild dogs when faced with crossing rivers in Kruger Park:

> The dogs swim well, but often show a disinclination to enter water, even when an antelope which they have pursued plunges into it. But there are times when a river, barring the way of a pack, must be crossed. Fearing crocs, the dogs may then gather on the bank and noisily announce their arrival. The great reptiles look upon the barking as a dinner bell, and hasten in the direction from which it comes. The dogs, having drawn all the crocodiles within hearing of them to one spot, career as fast as they can along the bank for a couple of hundred yards. Then they plunge into the stream, and cross it, before the crocs can checkmate their strategy.

Intelligent and wily, wild dogs often use man-made structures to assist them during a hunt. They will corner a fleeing antelope against the wall of a building, a raised walkway or the electrified fences found around many of the game reserves in South Africa.

The dogs are coursers – they do not stalk their prey but rather pursue and exhaust it. They are known to use the roads in reserves for easy access to areas and to gather speed during the chase. The packs in the Kruger are particularly fond of using the network of roads in the park instead of running through the long grass and the thick bush with its many obstacles. This seems to be particularly true in the summer months when the vegetation cover is taller and denser than during winter.

Opposite The Dwarsberg pack in Madikwe Game Reserve has thrived over the years, delighting tourists there. The size of the pack has varied from time to time, but averages around 15 dogs. This has made them a consistently successful hunting unit and allows them to dominate the smaller Collection pack when their paths cross.

This spread A growing awareness of these fascinating predators has inspired artists, ceramicists and even wine makers. The painter Andre de la Rosa and Ardmore Ceramics studio are but two of many who have created works with the dogs as a theme, while Borg Family Wines has produced a fine range of Painted Wolf wines.

Ancient Spirits

In our search for traditional knowledge on the wild dogs we found a smattering of folklore about them amongst the Bushmen in the Ghanzi area in north-west Botswana, but we looked still further back in time and discovered several intriguing references to wild dogs in Bushman mythology and in their rock art, which provides an extraordinary record of the daily life and spiritual beliefs of a Stone Age people from thousands of years ago.

The Bushmen, or San, are southern Africa's oldest inhabitants. Calling themseves the First People, they roamed the plains and sheltered in caves and overhangs of the subcontinent for tens of thousands of years. They were traditional hunter-gatherers and subsisted off the land in harmony with the natural world around them. Incredibly there are today a few remnant communities of these remarkable people, living mainly in the Kalahari sandveld regions of South Africa, Botswana and Namibia.

The Bushmen have always loved storytelling, music and dance, which they link to their spiritual beliefs, to their folklore and to their experiences of nature. The older generation, men and women, are particularly skilled in telling stories and delight in repeating them around the campfire at night, when in the flickering light they conjure up vivid images of various characters and events. Many of these stories have been handed down through the ages, the underlying theme in most tales being their belief that in ancient times all animals were once people like themselves.

According to one legend, this changed when their Great God moulded the different forms of man and beast in the flames of the Everlasting Fire and branded each animal with different markings, using irons heated amongst the coals.

The brown hyena, the last animal created, is associated with death. Even today in every set of divining objects that are still thrown by a traditional Bushman shaman before an important

Right Bushmen of the Giant Tortoise clan in the Ghanzi area of Botswana dance around the fire as they await the arrival of their shaman. The more traditional amongst them believe that, before a hunt, some shamans temporarily transform themselves into wild dogs.

IN SEARCH OF THE AFRICAN WILD DOG

decision or a hunt, there is one called a 'brown hyena', which brings bad luck and death if it falls upside down.

But there is a different myth about the origin of death that involves the wild dogs. In a tense confrontation between the Moon and the Hare, the Moon maintained that just as he died and was reborn when he waxed and waned, so too would Man. The Hare, grieving the death of his mother at the time, vehemently declared that this could not be so as his mother was well and truly dead and showed no sign of coming back. A bitter argument ensued during which the Moon slapped the Hare in the face and split his lip and the Hare scratched the Moon's face, causing great disfigurement. This so enraged the Moon that he declared the Hare would forever after be hunted and eaten by wild dogs and that Man would die and not be reborn. And so it is that the Hare has a split lip, the Moon is permanently scarred and Man is not immortal. It seems, however, that the wild dogs got the best of the bargain because they still pounce on a hare with alacrity whenever they see one, for a quick and tasty snack!

Another story tells of their god known as !*Kaggen*, who fell out with a group of his friends and decided to take revenge on them. Now !*Kaggen* was a great magician and trickster with the power to transform himself into anything he wished, to change men into animals and to bring them back from the dead. So with a flick of his wrist he changed a group of men into a pack of wild dogs and set them against his friends whom he transformed into giants. Who won the fight is not recorded, although the dogs probably acquitted themselves well.

The Bushman word for a wild dog in the northwest of Botswana is !///*haru* and some of the more traditional Bushmen living there still believe that, before a hunt, their shamans or medicine men sometimes transform themselves into wild dogs, considered to be nature's ultimate hunters. But prior to setting out, they wrap their feet up in strips of bark taken from the silver clusterleaf tree (*Terminalia sericea*) so as to leave no scent or tracks in the sand, which might lead other predators to their kill.

After they have hunted they change back into men, remove the bark from their feet and walk away, leaving only human footprints behind. (The silver clusterleaf is deemed sacred by many local people and is said to be a conduit through which they can communicate with the ancestral spirits.) The Bushmen's admiration of the wild dogs' hunting prowess extends to the belief that if they smear some of its body fluids under their feet they will become better hunters, infused with the same agility and boldness the dogs display during the chase.

The early Bushmen were prolific artists (it is generally believed today that a shaman in each clan was also the artist) and left behind thousands of rock paintings and engravings across southern Africa. Of these only a handful depict wild dogs, perhaps the most spectacular being a frieze in the Erongo mountains in Namibia. The scene portrays a pack of dogs hunting down two antelope and the stance and attitude of the buck and the individual dogs shows a keen eye for detail. The frieze is considered an exceptional piece of rock art, as wild dog portrayals are generally small and insignificant and usually positioned towards the outer edge of the painting. This is possibly because the Bushmen believed that the depiction of larger animals like the gemsbok, giraffe and particularly the eland (the latter considered to *be* !*Kaggen's* favourite animal and indeed his creation) would imbue their shamans with greater power during their trance or healing dance. They could more easily enter the spiritual world to take on the power of the spirits and chase away illness and misfortune.

After as many as 20 000 years Bushman rock art came to an abrupt end around 1900 when colonial expansion was at its height – both the Boers and the British aggressively pushing into what had hitherto been Bushman domain. Although the Bushmen of today no longer produce the art of their ancestors, or even identify with it, they still recognise their role in nature and show a reverence for wild animals and the environment. This is something that sadly many of us have ceased to do, as the plight of *Lycaon pictus* testifies.

Opposite Tswiigha, an 83-year-old shaman living near Ghanzi in western Botswana, shared his traditional knowledge on the wild dogs with us. He is one of only three medicine men in the area who still perform trance-dance healing.

TWO

Makanyane

(Tswana)

NORTHWEST BUSHVELD

Left Two elephants take a long drink at Tlou Dam after a hot day in Madikwe Game Reserve. The dam is a favourite with the reserve's elephants and seeing them here is almost guaranteed. Rather aptly *tlou* is the Tswana word for elephant.

Madikwe Game Reserve – the start of our wild dog odyssey – was new territory for us and proved to be quite an eye-opener. Some four hours from Johannesburg, it has the 'big five' and other great game viewing. It is also, together with the De Wildt Cheetah and Wildlife Centre in the Magaliesberg, a pioneer in wild dog reintroductions and the concept of a 'metapopulation' as a means to conserve these endangered animals. Indeed the dogs from De Wildt and Madikwe, through their offspring, have over the years provided new blood for the formation of packs in other protected areas throughout South Africa as well as in the Northern Tuli Game Reserve in eastern Botswana.

Of all the animals in Madikwe, it is the wild dogs that have become synonymous with the reserve.

Madikwe, situated in North West province, has been a huge conservation success story. It encompasses some 75 000 hectares of previously degraded land that had been used for cattle farming. But overgrazing in this arid area, among other marginal agricultural practices, had left the land in such poor condition it seemed to offer very little economic benefit to anybody.

Studies showed that turning the land over to wildlife was the best option and, common purpose prevailing, the reserve was proclaimed in 1991. It is now one of the country's premier game reserves, representing a unique partnership between the North West Parks and Tourism Board, local communities and the private sector. Central to this three-way alliance is people-based wildlife conservation, whereby local communities gain financial benefits from conservation and private ecotourism. This approach has ensured support for the reserve and its activities from its immediate neighbours and has become a blueprint for long-term conservation across southern Africa.

Apart from this pioneering approach, Madikwe is also known for what was the world's largest game relocation – a massive undertaking appropriately called Operation Phoenix. After a phase of intense environmental restoration, which saw the eradication of alien plants, the reclamation of eroded areas and the removal

Opposite Wild dogs hunt by sight rather than by sound or smell so, as night closes in, they are unlikely to still be on the lookout for potential prey. They are, however, known to hunt on moonlit nights when – as luck would have it – their main prey species, impala, also tend to be more active than on darker nights.

Above and top In the late 1800s a road was built to link the Transvaal and Bechuanaland Protectorate (now Botswana). At one time the British South Africa Police used camels to traverse the distance, which was a busy coach route for passengers, traders and fortune-seekers heading further north.

of fences and dilapidated farm buildings, the reserve became home to a staggering 8 200 animals of 27 species that were re-introduced over a period of seven years. Most of the animals were sourced from reserves and breeding centres in southern Africa. However, some – including 50 disease-free buffaloes – came from zoos in the USA and the Czech Republic. Once the initial restocking was completed, like the legendary phoenix Madikwe rose from the ashes of depleted farmland to offer range to a multitude of game including all the big five, herds of many different antelope, zebra, giraffe, brown and spotted hyena, cheetah and wild dog.

The reserve is located in a transition zone of Lowveld, mixed bushveld and Kalahari thornveld ecosystems, the resultant diversity of habitats being a major part of its attraction. It is scenically beautiful with rocky outcrops, wooded thickets and sweeping plains like nowhere else in South Africa that slope gently towards its main water course, the Marico River. The imposing Dwarsberg hills dominate the southern region, rising some 200 metres above the surrounding bush. A series of ridges known as the Rant van Tweedepoort divides the reserve into two roughly equal parts. Those areas that were previously farmed and cleared of alien trees have remained open and now offer excellent game viewing, emulating the Serengeti experience.

Several important archaeological and historical sites are located within the reserve. Early Stone Age artefacts dating back around a million years have been found along the Marico River, while a number of Iron Age sites are located along the Rant van Tweedepoort and the Dwarsberg. Tswana people settled in the region in the 1300s and things were relatively peaceful for the next 500 years. Then, from around 1800, increasing numbers of missionaries, adventurers and hunters arrived, stayed or moved further north. One such was Dr David Livingstone, whose famous travels started out from the nearby Moffat mission station at Kuruman.

Of far greater significance to the course of history, however, was the breakaway Zulu chief Mzilikazi, who plundered this area for some years. He was eventually routed by the Voortrekker vanguard, with their superior resources (horses and gunpowder), fleeing into western Zimbabwe. There he established himself as the first king of the Matabele nation in the late 1830s. The vacuum left by his 'path of blood' was soon filled by Voortrekkers, who began to establish farms and then settlements in the vicinity. A wagon road was established, linking the Transvaal Republic with the Bechuanaland Protectorate (now Botswana). This became a regular couch route for passengers and goods deeper into the southern African interior, as well as for a steady stream of traders and fortune-seekers first to the diamond fields and later gold fields. It became known as the Zeederberg Trail, after the service that at some stage famously used zebras in place of mules and horses to pull its coaches.

The Mafikeng Road connection

Perhaps one of South Africa's greatest literary figures, Herman Charles Bosman, was posted to the Marico district as a young teacher at the age of 21 in 1925. His stories, most notably *The Road to Mafeking*, are filled with references to the Dwarsberg, Abjaterskop, Derdepoort and Nietverdiend – all familiar landmarks to anybody who visits Madikwe. Bosman stayed in the area for only six months, but such was his experience there it provided him with the setting for many of the short stories for which he later became famous.

As luck would have it, our game ranger on several trips to Madikwe was Patience Bugatsu of Thakadu River Camp. She is an avid Herman Charles Bosman fan, having studied his works while at the convent school in Molatedi village, just outside the reserve. We were blown away by her literary skills, but things got even better when we realised that Patience had a passion for the bush and an unerring instinct for tracking wild dogs, which proved extremely useful as the two packs in the reserve – known as the Collection and the Dwarsberg packs – are like their cousins elsewhere: extremely elusive. (The Collection pack is named after the Madikwe Collection, a group of lodges that sponsored the translocation of five female dogs from Shamwari Game Reserve in the Eastern Cape to Madikwe in 2007.) Patience seemed to feel their presence in the long grass or in a thicket, but then she uses her sense of smell far more than most people: one thing about wild dogs is their smell – feral and strong!

Patience's grandfather had been a respected traditional healer in the district; his reputation opened doors for us as we went

Above A young female from the Dwarsberg pack rests up in the long grass in Madikwe. The pack has a sizeable home range: while we spotted them in the western part of the reserve, they were often seen in the south and east as well. The pack is named after the Dwarsberg range, a dominant feature of the area.

looking for Tswana folklore on the wild dogs. And so it was that, with Patience as interpreter, we sat cross-legged on a grass mat in Emmanuel Pele's consulting room. Only 25 years old and recently graduated as a sangoma, or diviner, he already had considerable knowledge of the use of various animal parts and plant products in the treatment of illness, the interpretation of misfortune and for fortune-telling. Called to his profession by the ancestors, he utilises different methods of mediation with the spirit world to help members of his community who consult him about their problems.

Part of his divining method is to 'throw the bones', and it seemed appropriate at this point to ask him whether the wild dog played any part in his divinations, or in the concoction of his traditional medicine, or muthi. *To our amazement he shook his head, saying that he had never seen or heard of the animal before. Seriously thrown by this admission, we showed him a photograph we had recently taken of a group of wild dogs and waited in silence as he studied the image. After several minutes, Emmanuel announced that he noticed that they were pack animals and was therefore confident that their scats, placed at the entrance to a*

Opposite Storm clouds gather over Madikwe, cross-lit by the late afternoon light. Summer thunderstorms in the reserve can be impressive and even fear-evoking as they roll over the horizon and rumble towards you across the veld.

Right Ranger Patience Bugatsu has almost a sixth sense when it comes to finding wild dogs, which was hugely helpful as the two packs in Madikwe proved to be very elusive.

cattle kraal, would prevent the herd from dispersing when let out to graze. Greatly impressed with his quick improvisation, we placed his consulting fee on the ground at his feet – the usual way to appease the ancestral spirits – and headed back to our vehicle.

We thought, at first, it might have been Emmanuel's youth that accounted for his ignorance of these creatures but, after several visits to other sangomas in the area, it became obvious that any traditional knowledge of wild dogs had long since been forgotten. It was hugely sobering to realise the extent to which they had so completely disappeared from the lives of rural people in the area.

Over the coming months, on our various trips to the region, we cast our net much wider for any myths, legends or stories – anything – that may have been handed down through the generations about these animals. We managed to collect various titbits of information that were really intriguing, but no folklore as such. The Tswana word for wild dogs in North West province is makanyane, *which alludes to their agility during a hunt. The word is sometimes used to describe petty criminals and pickpockets in the local townships. In Botswana, a mere 20 kilometres to the north of Madikwe, the word* lekanyane (the singular form of makanyane) *is used to describe a soccer player, more specifically a goalkeeper, who is particularly agile. The Batswana defence force trains to a song about the wild dogs: 'Lo kile lwa se bona kae makanyane a aja pitse?' – 'Have you ever seen wild dogs preying on a horse?' The meaning of this escaped us until we were told the soldiers refer to themselves as horses, which the dogs might well hesitate to attack as they could be on the receiving end of a lethal kick.*

Historically, wild dogs occurred with other predators in this northwestern area of the country. However, prior to their re-introduction into Madikwe they were encountered only on very rare occasions in recent times, and then just as small itinerant groups that had wandered in from Botswana. The reserve's long-awaited wild dog programme was launched in December 1994 when three captured male dogs from the Kruger National Park were released into a holding enclosure in Madikwe. They were joined a few months later by three captive-bred females from the Sasol-sponsored pack at De Wildt Cheetah and Wildlife Centre, a private breeding facility that has played a pivotal role in the Madikwe wild dog story. A unique opportunity presented itself here, to closely monitor the interaction between the wild and captive dogs and observe the formation of a new pack under the artificial conditions in the boma. Dr Markus Hofmeyr, veterinarian and field ecologist in the reserve at the time, embarked on an emotional roller-coaster ride when he began tracking the fate of the first wild dog pack in Madikwe.

In July 1995, after some six months of acclimatisation in the holding boma, during which time an alpha pair emerged and a pack hierarchy was established, the dogs were released into the reserve. Five of the six had either radio collars attached or radio transmitters implanted: North West Parks was taking no chances on losing its new stars! After their release they were monitored by a combination of telemetry readings and visual sightings by Markus and his team, and over the coming months the dynamics of the pack were observed and recorded.

Earlier concerns that the captive-bred females might not be able to hunt were soon laid to rest; within weeks they were playing an active role in the chase and kills. It is likely that the skills of the wild-caught males rubbed off to some degree on the females, but it was evident that their ability to hunt was so ingrained it had not been lost. Interaction between the females and other predators was also a concern, but this too proved groundless after the pack survived at least three observed encounters with their biggest threat – lions. On all three occasions the lions came from areas where they had not previously encountered wild dogs, and their not recognising and eliminating potential competitors (as they are generally wont to do) could perhaps have contributed to the dogs' survival.

The dogs had clearly gone on to successfully operate as a cohesive pack, had adapted to their new environment, coped with the pressures of competing predators in the reserve and produced several litters of pups. By September 1997 they numbered a very healthy 23, but then disaster struck. The alpha male disappeared and for several days Markus and his researchers could find no trace of him. But as the rest of the pack seemed unwilling to move far from their den, it became obvious he had died inside the burrow. It took several more days before the body could be retrieved, by which stage it was so decomposed an autopsy could not be done.

Four days later Markus came upon the pack in a piteous state: the pups were weak and dejected and the rest of the dogs lethargic and indifferent to their predicament. Markus's heart sank: one of his greatest concerns had been the possibility of rabies or distemper breaking out amongst the dogs, as these diseases had been known to wipe out entire packs. His worst fears were realised after a post mortem on one of the dogs confirmed the dog had died of rabies. Most likely the alpha male had killed or been bitten by a rabid jackal and then passed the disease on to the others through the exchange of saliva, either through the regurgitation of meat or through social licking and grooming.

The start-up pack was eventually reduced to a paltry three dogs. In spite of the setback, though, Markus was determined to continue with the wild dog project. De Wildt was able to offer a further two wild-caught males and three captive-bred females for release, and Sasol (together with Tau Lodge in the park) was prepared to sponsor another relocation of dogs to the reserve. This time round, however, management took the decision to inject the animals against rabies on an annual basis, as opposed to the once-off inoculation the first dogs had received. This proved to be the answer to this and all subsequent introductions: when an outbreak of rabies swept the region in 2000, dogs that had been vaccinated annually remained resistant to the disease.

Looking back, Markus considers the release of the first wild dogs into Madikwe to have been an unqualified success. The learning curve led to a new model and subsequent successful relocations into Madikwe, as well as other protected areas. Important lessons learnt in the process have proved to be invaluable: to regularly monitor the dogs in the boma and after their release to record their progress and keep track of their circumstances; to inoculate against rabies on an annual basis where the disease is prevalent; to manage any dispersal groups and ensure the introduction of new blood lines; and to maintain predator-proof perimeter fencing around smaller protected areas to prevent the dogs from escaping into surrounding lands (which they are particularly skilled at doing).

Opposite This lioness gave a warning growl as we approached. Her threat was unmistakable and we understood why when we spotted several cubs playing close by. We did wonder though what chance a wild dog pack would have against her.

Less ordinary lives

Markus left Madikwe in 1999 to join the Veterinary Wildlife Services in the Kruger National Park. The current field ecologist is also a Hofmeyr and, although only distantly related, a commitment to conservation is obviously a family trait. Declan Hofmeyr's involvement with the wild dogs in the reserve spans some seven years and he's found that managing a species that is part of a metapopulation requires loads of lateral thinking!

Declan was a huge help to us, determined that we got the best photographs and 'the whole story'. After several weeks' absence from the reserve, we received a call from him to say the Collection pack had denned in an old aardvark hole that was close to a road and easily accessible. But, between the time we heard from him and our arrival a few days later, the pack was almost completely wiped out in a confrontation with lions. Only three dogs remained, the alpha female so badly injured she was unable to feed her pups, all of which subsequently died. In an attempt to salvage something from the disaster Declan took the female, who had a tenuous hold on life, to a vet in Thabazimbi, some 100 kilometres away. After numerous operations to repair her throat and neck, she somehow pulled through. He collected her several weeks later and placed her, still anaesthetised from the journey, in one of the reserve's bomas.

Meanwhile a female dog was available at the neighbouring Pilanesberg National Park for relocation and it seemed an ideal opportunity to integrate her into what remained of the pack. But Declan faced a dilemma: no adult dog had previously been successfully introduced into an existing pack of both sexes. The hierarchy being such that an unrelated dog is typically turned upon and killed (the females are apparently particularly vicious during introductions, and are more likely than the males to fight to the death). But the Collection pack, he reasoned, had been torn apart by the lion attack and the alpha female was still in a compromised state. So, after consultation with other ecologists, he decided 'to use a combination of science, medicine, wild dog ecology, timing and good old fashioned dumb luck'. That same afternoon he drove to Pilanesberg, darted and captured the female and put her, also still sedated, into the boma.

As a precaution he thoroughly doused both dogs and the area around them with a drug called DAP – dog appeasing pheromone, which is used on domestic dogs and cats to reduce aggression. (Apparently an appeasing pheromone is produced in the ear canals of both the alpha male and alpha female of a wild dog pack, especially during the breeding season, which has the effect of calming the other dogs.) The following morning he approached the area with some trepidation and found to his relief that both dogs were still alive, although minor bite marks around their faces attested to some interaction during the early hours. He was elated.

He then moved onto the next phase of his rescue plan, which was about the time we happened upon him in the southeastern part of the reserve. The two other dogs (a male and female) that had survived the lion attack had remained in the vicinity of their den. The idea was to capture them and put them in the boma with the two females, so they could bond and form a new pack. We watched in amazement as Declan called the dogs to his truck by lifting and dropping the tailgate, the sound produced being part of the dogs' Pavlovian conditioning that signalled the arrival of food. He had instilled this response into them when they'd first come to Madikwe years before, so that he could call them in if the need arose, as it now did. Lured out of the dense thicket by the sound of the tailgate, and no doubt the smell of the freshly killed impala that had been attached to a tree, the dogs were easily darted and taken to the boma.

While they were asleep there he injected all four dogs with an anti-psychotic drug that would cause them to be slightly euphoric for about three days, then rubbed them against one another and smeared them with one another's body fluids to make them all smell the same, like a natural pack would. As before, he liberally doused the dogs and their surroundings with DAP and left them to wake up, lying side by side, in the dark. Once again when Declan approached the boma the next morning, the familiar feeling of anxiety knotted his stomach, but four dogs stared back at him, albeit slightly unsteady from their cocktail of drugs, all still alive and with no evidence of fighting. Over the following weeks they continued to bond and established a pack hierarchy, the alpha female incredibly managing to keep her position in spite of her slightly weakened condition. All was going according to plan and, with their release imminent, Declan was looking forward to seeing them running free in their natural environment.

This page After the Collection pack was almost completely annihilated in an altercation with lions, Madikwe field ecologist Declan Hofmeyr embarked on a rescue plan to try and salvage what was left of them. To capture the two surviving dogs that remained in the vicinity of their den, he enticed them out of the thick bush with the smell of a freshly killed impala, which he tied to a tree.

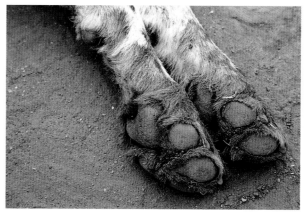

IN SEARCH OF THE AFRICAN WILD DOG

This spread Capturing wild animals is always a stressful affair, but much to everybody's relief the two dogs were easily darted and quickly succumbed to the tranquilisers. Still, it was unsettling to see them stretched out on the sand, their tongues lolling out and pools of saliva forming under their chins, so vulnerable. Working quickly, Declan took the opportunity to do blood tests, all the while making sure the dogs were comfortable. The whole procedure lasted less than an hour.

Above Professor Henk Bertschinger and Ann Van Dyk draw blood from a tranquilised dog during routine testing at De Wildt Cheetah and Wildlife Centre. Ann is passionate and determined about the long-term survival of both cheetahs and wild dogs and has spent her adult life working with these two species.

But at the eleventh hour things, incredibly, went horribly wrong. The day before their release the Dwarsberg pack, which had routed the Collection pack on at least two previous occasions, happened past the boma. In the ensuing fracas through the fence, two of the four dogs inside the enclosure were killed and the other two badly bitten. Against all odds the alpha female survived the fight, but needed more stitching to her recently healed neck and throat. Totally gutted by the experience, Declan acknowledged it had been the most heart-wrenching day he had ever had to endure as field ecologist. 'It seems,' he said with an air of resignation, 'that wild dogs are destined to live lives less ordinary than others.'

Most of Madikwe's wildlife has thrived over the years and many species have increased significantly in number since they were introduced. If they are to be viable in the long term, however, then the big need now is for more land. With this in mind plans are afoot to link up with Madikwe's sister reserve, Pilanesberg, via a corridor between the two. The bigger picture is to expand towards the Waterberg in the east and even possibly into neighbouring Botswana to the north and west. This is good news indeed for the wild dogs, with their requirement of vast home ranges.

De Wildt dream realised

The story of Madikwe's wild dogs is closely linked to that of the De Wildt Cheetah and Wildlife Trust, the brainchild of a remarkable woman. Ann Van Dyk has spent a lifetime in cheetah conservation and, more recently, has turned her attention and considerable expertise to another of Africa's endangered carnivores – the wild dog. From humble beginnings De Wildt has grown into a significant authority on both these predators and conducts ongoing breeding, research, releases, education and community outreach programmes that will ultimately enhance their chances of long-term survival.

But these activities cost money, lots of it, which has made the sponsorship received from Sasol towards various wild dog projects invaluable. First granted in 1994, the sponsorship has enabled De Wildt to build an enclosure for captive-bred wild dogs. Since its completion nearly 100 000 tourists have passed through the camp, as part of a guided tour of the facility. The educational value of this is obvious; more especially so because many visitors, a good percentage of them school children, had not previously seen a wild dog. Quarantine camps were completed in 1996 so De Wildt could test any wild dogs sent to the centre for possible diseases, the species being highly susceptible to those that affect domestic dogs. A few years later a further six enclosures were built to ensure the blood lines of captive-bred dogs remained separate to prevent weakened DNA that results from inbreeding. There are currently eight different breeding lines in the Sasol/De Wildt camps and Ann is confident that, should they need to supply any dogs to game reserves, the animals will be genetically sound.

Professor Henk Bertschinger, head of the Wildlife Unit at the Faculty of Veterinary Science at Onderstepoort in Pretoria, has long been associated with the De Wildt Cheetah and Wildlife Centre and is involved in various research programmes, in conjunction with the University of Pretoria and with the approval of the university's Ethics Committee. Sasol has set great store by Professor Bertschinger's work and has funded research into rabies in wild dogs, DNA mapping to establish the heterozygous state of the De Wildt dogs, distemper trials on the efficacy of 'live' and 'dead' vaccines, contraception and the freezing of wild dog semen.

The successful release of captive-born dogs into the wild has always been close to Ann's heart. She was able to realise her dream in 1995 when three females born in one of her camps were bonded with three wild-caught males from the Kruger National Park and released into Madikwe (their story recounted earlier). Despite some scepticism prior to the project, which was considered ground-breaking at the time, the dogs survived and thrived, the pack growing to more than 20. The success of the undertaking paved the way for more of its kind, giving momentum to De Wildt's plans to implement a national programme for the conservation of free-roaming wild dogs and their ecosystems – much as they have done with cheetahs. The idea is to offer landowners an alternative to shooting wild dogs that come onto their property, by giving them the option of contacting De Wildt. The centre will then send out a team to capture the dogs and temporarily house them until they can be relocated to a suitable protected area.

Bumpy ride to Pilanesberg

Pilanesberg National Park (a somewhat misnamed provincial reserve) is situated in the crater of an ancient caldera, or collapsed volcano. Long extinct, its eruptions some 1,2-billion years ago resulted in an 'alkaline ring complex', the kind of which occurs in only two other volcanoes in the world, one in Russia and the other in Greenland. The volcanic origin of the area, combined with erosion of the crater, has led to a fascinating setting for this 55 000-hectare park, which traverses rugged terrain, forested ravines, bushveld and rolling grasslands.

Established in the apartheid-era Bophutatswana homeland in 1979 Pilanesberg, like Madikwe some time later, went through a similar metamorphosis: transformed from an area that had been severely degraded by marginal farming to one that has been all but restored to a natural state. Here it was thanks to Operation Genesis which saw a veritable Noah's Ark of animals released into the park.

Pilanesberg is today home to a great variety of wildlife, including the big five and several rare and endangered species including black and white rhino, sable antelope and wild dog. After the success of the wild dogs in Madikwe, North West Parks decided to introduce a pack into Pilanesberg, not only for their ecotourism value but also because it would be yet another area that might be able to contribute towards the establishment of a viable metapopulation.

Gus van Dyk, who had played a pivotal role in the release of dogs in Madikwe, as well as a series of successful introductions of lions, cheetahs and spotted hyenas, took control of the operation. A pack of nine dogs duly found a home in the park in June 1999, after several months of bonding in a boma. This reintroduction, together with others over the years, has ensured that the dogs have become an important attraction for tourists in the reserve and a vital link in the chain for the future preservation of the species.

Gus left Pilanesberg in 2006 to join the private Tswalu Kalahari Game Reserve in the Northern Cape's Kalahari sandveld area near Kuruman and handed over to Steve Dell, the current field ecologist in the park. He is an equally avid proponent of the metapopulation concept and firmly believes they are a vital part of a healthy natural ecosystem.

This spread Wild dogs try to avoid areas frequented by lions as any confrontation between them will almost certainly result in death or injury to the dogs. This avoidance might be a factor in determining when dogs choose to hunt, leaving the hours of darkness to the big cats. The dogs do occasionally hunt at night, but only when there is bright moonlight.

Makanyane ~ Sotho/Tswana

IN SEARCH OF THE AFRICAN WILD DOG

This spread Brown hyenas are solitary and nocturnal and not often seen during the day. It was a great treat, therefore, when we saw this one not only come into full view, but stumble into a group of dogs, which lost no time in chasing it off.

The Bushmen believe the brown hyena was the last animal created by the gods. It is associated with death and every set of divining objects thrown before an important decision or hunt contains one called a 'brown hyena', which brings bad luck if it falls upside down.

MAKANYANE ~ SOTHO/TSWANA

In search of the African Wild Dog

Above In spite of its size this elephant bull was almost hidden behind some reeds that lined a section of the Marico River on the eastern boundary of the reserve. The name Marico comes from the Tswana word *madiko* and means 'there is blood' – an allusion to an earlier time of conflict here.

Above An adult kudu bull has magnificent horns, which will twist three full turns by the time it is fully grown. While lions and spotted hyenas will bring one down, a big bull will generally prove too large for wild dogs, which take kudu calves and sub-adults instead.

Steve was as determined as Declan had been that we get our story. Knowing we were working in the area, he invited us over to observe the capture of three female dogs, which were to be transported by air to the Hluhluwe Imfolozi Park in KwaZulu-Natal. We grabbed the chance as the operation could make for great action shots – which it did. This was a combined effort between the two conservation authorities – pilot Greg Nanni and veterinarian Dave Cooper from Ezemvelo KZN Wildlife, and Steve and his team from North West Parks.

As happens on these occasions, things were a little stiff at first, but soon improved as the morning progressed. Perhaps the thaw started when Steve whipped the tarpaulin off the back of his truck and everybody got the first whiff of the zebra carcass that was to be used to entice the dogs to a spot in the boma so that Dave could get a chance to dart the three females. An old hand at game

This spread The relocation of three female dogs from Pilanesberg in North West province to Hluhluwe Imfolozi in KwaZulu-Natal offered great story and photo opportunities. Darting the three proved to be a tense procedure as the task needed to be completed quickly, before they had eaten their fill of a zebra carcass and moved off.

capture, the night before he had indulged in both garlic snails and a garlic steak to help disguise the bouquet he knew was to come. There was no doubt it helped him, as he seemed impervious to the lot, but for the rest of us it was an eye-watering affair.

Slightly hampered by the drizzly weather that had moved in overnight, Dave loaded his dart gun with the necessary immobilising drugs while someone else heaved open the heavy gates to the boma. Steve drove his truck towards the excited dogs that were twittering frenziedly, the legs of the zebra wobbling about as several men clung to the carcass prior to throwing it off the back. A fairly tense time followed as Dave worked fast but accurately to dart the females and then conduct various tests. The three were laid out on the rear of the truck and, as we looked at them, their fur matted and damp from the drizzle, it struck us anew as to how small they were, their hindquarters and long legs quite puny for such efficient hunters. What also hit us though was the smell, which wafted up in waves from the unconscious dogs, their reputation in this respect being well deserved.

From there we travelled in convoy to the Pilanesberg airport, some 20 kilometres outside the park, and watched Greg organise the loading of his passengers into a small Cessna. The three wild

dogs took up every centimetre of space behind the front seats. The day had started to warm up and Greg was anxious to leave quickly to try and avoid too much turbulence on the three-and-a-half hour flight back to Hluhluwe Imfolozi. Even before the doors were closed the stench inside was like a solid mass.

Some weeks later we contacted Greg to find out how it had gone. The trip had been pretty bumpy, he said, and with the sun pouring in through the windscreen – adding to the reek of garlic from Dave – the dreadful odour from the dogs in the back had almost finished him off. Not only had it been their fetid smell, but they had belched and broken wind all the way to KwaZulu-Natal, the appalling stink of rotten zebra filling the cabin like a toxic cloud. To add to his woes, he'd developed a crick in his neck from continually straining towards the air vent above his head. It had taken him days and many showers to get rid of the stench that had permeated every pore of his being. Dave, he said after a pause, had sailed through it all without so much as a twitch. However, within minutes of his walking through the front door, his wife and fox terriers had exited out the back and not even his horses had wanted anything to do with him!

IN SEARCH OF THE AFRICAN WILD DOG

Above and opposite The tranquilised dogs were loaded up and transferred into the back of a light airplane for a three-hour flight to Hluhluwe Imfolozi. They survived the trip with no noticeable ill effects; they were duly placed in a holding boma subsequent to being released into their new home several months hence.

MAKANYANE ~ SOTHO/TSWANA

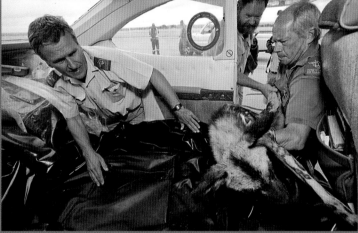

Overleaf Wild dogs hunt in silence, communication between members being by sight and smell. Care for their young extends even after they are old enough to join in the hunt, when the adults will stand back from a carcass to allow them to eat their fill.

What is the Future of Wild Dogs in North West Province?

Professor Henk Bertschinger of De Wildt Cheetah and Wildlife Trust comments:

Wild dogs are fascinating predators, their entire social behaviour is geared towards maximising their chances of survival and there is no other carnivore with a reproductive drive that comes close to theirs. The greatest threats to their survival are humans, the diseases rabies and distemper, and large predators such as lions and spotted hyenas. Wild dogs need space, with reserves of less than 35 000 hectares being too small to permanently contain a pack. They breed up very quickly, which leads to break-outs into neighbouring properties where they are likely to be persecuted. This means that, particularly on smaller properties, population numbers have to be contained. Reserves that carry high densities of other large predators are also unsuitable.

The metabolic rate of wild dogs is exceptionally high (greater than that of a border collie), which means that for the most part they have to eat on a daily basis. Kleptoparasitism (theft of their kills by other predators) forces them to leave an area – if not, they will die of starvation. In areas where rabies and distemper are a problem, the diseases can be controlled by vaccinations using drop-out darts. The bottom line is, given sufficient suitable habitat, the wild dog will survive.

Where to See Wild Dogs in North West Province

The international code, if calling from outside South Africa, is +27, drop the first 0.

Madikwe Game Reserve
Tel: +27(0)18-350-9931/2 / Fax: +27(0)18-350-9933
E-mail: madikweadmin@telkomsa.net
Website: www.madikwe-game-reserve.co.za / www.tourismnorthwest.co.za

The Madikwe Collection
Tel: +27(0)11-805-9995 / Fax: +27(0)11-805-0687
E-mail: reservations@madikwecollection.com
Website: www.madikwecollection.com
These reservation details pertain to the following Madikwe lodges:
Tuningi Safari Lodge, Motswiri Private Safari Lodge, Buffalo Ridge Safari Lodge, Thakadu River Camp, Madikwe Farm House, The Bush House, Madikwe Hills Private Game Lodge and Morukuru Lodge.

Etali Safari Lodge
Tel: +27(0)12-346-0124 / Fax: +27(0)12-346-0163
E-mail: info@etalisafari.co.za
Website: www.etalisafari.co.za

Impodimo Game Lodge
Tel: +27(0)83-561-5355 / Fax: +27(0)83-561-0653
E-mail: steve@impodimo.com
Website: www.impodimo.com

Jaci's Lodges
Tel: +27(0)14-778-9900/1 / Fax: +27(0)14-778-9901
E-mail: jaci@madikwe.com
Website: www.madikwe.com

Madikwe Farm House
Tel: +27(0)11-315-6194 / Fax: +27(0)11-805-0687
E-mail: info@madikwesafaris.co.za
Website: www.madikwefarmhouse.com

Madikwe River Lodge
Tel: +27(0)14-778-9000 / Fax: +27(0)14-778-9020
E-mail: lodge@madikweriverlodge.com
Website: ww.madikweriverlodge.com

Madikwe Safari Lodge
Tel: +27(0)11-809-4300 / Fax: +27(0)11-809-4400
E-mail: safaris@ccafrica.com
Website: www.ccafrica.com.

Makanyane Safari Lodge
Tel: +27(0)14-778-9600 / Fax: +27(0)14-778-9611
E-mail: enquiries@makanyane.com
Website: www.makanyane.com

Mateya Safari Lodge
Tel: +27(0)14-778-9200 / Fax: +27(0)14-778-9201
E-mail: reservations@mateyasafari.com
Website www.mateyasafari.com

Mosetlha Bush Camp
Tel/ Fax: +27(0)11-444-9345
E-mail: info@thebushcamp.com
Website: www.thebushcamp.com

Rhulani Safari Lodge
Tel: +27(0)11-469-5082 / Fax: +27(0)11-469-5086
E-mail: rhulani@madikwegamereserve.com
Website: www.madikwegamereserve.com

Royal Madikwe Lodge
Tel: +27(0)82-787-1314 / Fax: +27(0)86-671-6125
E-mail: reservations@royalmadikwe.com
Website: www.royalmadikwe.com

Tau Game Lodge
Tel: +27(0)11-314-4350/49 / Fax: +27(0)11-314-1162
E-mail: taugame@mweb.co.za
Website: www.taugamelodge.co.za

THREE

Amakentshane

(Zulu)

ZULULAND

Left The open grasslands and savanna woodlands in the south of Hluhluwe Imfolozi Park are excellent for game viewing and a good place to see white rhino. During the 1960s the park became world-renowned for saving the southern African population of this species from extinction.

We cut our teeth on wildlife photography in Hluhluwe Imfolozi Park in northern KwaZulu-Natal and, as with any place which holds special memories, it's always great to return. This time we were on foot and heading towards a wild dog den that researchers with the KZN Wild Dog Project had been monitoring for weeks. As we squelched our way through the coarse, wet sand of the Umfolozi River, intent on the tracks in front of us and wary of the buffalo bull in the reeds to our right, a warning bark rang out from the riverbank a short distance ahead. It was deeply moving to be barked at by a wild dog, being such a rare creature, and we stood breathless for a moment as we savoured the privilege.

There are only a handful of protected areas in KwaZulu-Natal, all in the north, which have wild dog populations of any significance. They include Hluhluwe Imfolozi Park and its close neighbours Mkhuze Game Reserve and Thanda Private Game Reserve. But there are other reserves in the vicinity looking to include dogs in their line-up of animals with a view to help conserve the species and to attract tourists. Among them, Mun-ya-wana Game Reserve (which includes Phinda Resource Reserve) and the Zululand Rhino Reserve, which are well known for their sound conservation records. The increased interest in wild dogs in the area is good news indeed, as more core populations will mean more dogs becoming available to disperse and form new packs throughout the region.

Founded as two separate reserves back in 1895, Hluhluwe Imfolozi Park is the largest game reserve in KwaZulu-Natal and one of the oldest in Africa. It is located in an area steeped in Zulu history and there is a tangible sense of the past as one travels though the park. Indeed, it was a mere 25 kilometres to the west of Imfolozi that the mighty Zulu nation stirred to life under the chieftaincy of King Shaka. In 1816 he assumed the leadership of what was then a rather insignificant Nguni clan and one of many in the vicinity. Conflict between the various groups was commonplace, largely because land for grazing cattle was increasingly in short supply. Shaka led his people in a series of bloody battles against rival clans and within the space of 12 years defeated and united most of them under his rule. The area

Opposite While every wild dog has its own unique colour pattern, the rounded ears and broad muzzle are almost always dark. The hair is coarse and short and the skin has numerous scent glands that are responsible for the pungent odour so characteristic of the species.

that Hluhluwe Imfolozi Park now covers was considered to be Shaka's royal hunting ground: a series of depressions in the southeastern corner are the remains of several huge game pits dug by his warriors for a massive hunt to celebrate his victory over Chief Zwide, a long-standing adversary.

The park scenery varies greatly and ranges from steep forested hills with grassy slopes and riverine woodland in the Hluhluwe section to the flatter and drier savanna woodland and open grassland in the southerly Imfolozi section.

Few of the rivers contain permanent water and even the Hluhluwe, as well as the Black and White Umfolozi rivers, stop flowing during very dry periods. This varied landscape is home to herds of antelope, giraffe, wildebeest, zebra, elephant, buffalo and groups of rhinos. Predators include lion, leopard, cheetah, spotted hyena and several packs of wild dogs – the largest number of dogs in South Africa outside Kruger Park.

The reserve became world-renowned in the early 1960s as the place of Operation Rhino, a campaign mounted in Imfolozi to save the white rhino from extinction. So successful was this, it soon became possible to relocate surplus numbers of these animals to other protected areas throughout southern Africa and subsequently the continent. It is said that today almost every white rhino across Africa can trace its origins back to this reserve. Wouldn't it be fantastic if something similar could be done for the wild dog?

During the late 1800s it was apparently not unusual to come across packs of wild dogs while traversing KwaZulu-Natal. Frederick Vaughan Kirby, in his book *In Haunts of Wild Game*, a journal of his wanderings between 'Khalamba' and 'Libombo' (published in 1896) observes:

> These dogs hunt in packs, by day or night, of from ten to twenty in number. They exhibit the greatest boldness in attack, and will frequently follow a solitary horseman for miles. What their object may be it is hard to say, as I have never known a case of a horseman being attacked; but they have attacked men on foot.
>
> Only a few months ago one of a pack of fifteen or sixteen took a duiker which I had shot almost from under my feet as I was about to pick it up, and I had to relinquish it to them to enable me to open fire on them, as their actions were unpleasantly demonstrative. Not until I had dropped four of their number on the spot did they make off…

Top Frederick Kirby looked every inch the swashbuckling hunter so beloved by society and the press in the late 1800s and early 1900s.

Right While a buffalo herd can be quite placid, lone bulls have a reputation of being bad-tempered and unpredictable. Old and arthritic, they sometimes hang out in small groups, which gives them some protection against lions.

Opposite A white rhino's huge bulk saves it from most large predators, but it has few defences against man. Its horn is prized in the Far East as an aphrodisiac and for traditional medicine, while in the Middle East it is used for ceremonial dagger handles. It is targeted by poachers, often working for crime syndicates, even in protected areas.

Before the reintroduction of 22 wild dogs to Hluhluwe Imfolozi between 1980 and 1981, the species had not been seen in the area for some 50 years. Systematically exterminated as vermin by farmers and game rangers alike, the last few individuals were shot or poisoned in about 1933. Despite a fluctuation in numbers through the years, their release into the park has been a success, a series of additional reintroductions further increasing the population and stimulating breeding. Ezemvelo KZN Wildlife has found (as has been the experience elsewhere) the most successful reintroductions of wild dogs are those where the dogs are first bonded in a boma and released only after they have formed a socially integrated and cohesive pack.

Today there are an estimated 83 wild dogs in about 10 packs in Hluhluwe Imfolozi and their continued presence has prompted the formation of the KZN Wild Dog Project. The project – the brainchild of The Endangered Wildlife Trust, Ezemvelo KZN Wildlife and the Smithsonian Institute in the USA – involves an extensive community awareness campaign with the park's immediate neighbours, with a view to expanding the dogs' home ranges into those areas. By informing people about wild dog behaviour, their distribution and their potential impact, the project hopes to turn around the perception that it is a 'problem animal'.

Keeping the gene pool clean

In attempts to change attitudes towards the dogs and create a safe environment for them, the process of developing relationships with the surrounding game ranchers, landowners and local communities has been slow but steady.

Resolving potential conflicts is central to the success of the project and Ezemvelo KZN Wildlife and The Endangered Wildlife Trust have responded quickly to situations when and wherever they have occurred. Most concerns have been about farmers losing stock to the wild dogs and being compensated for any losses. To date follow-ups to claims of wild dog attacks on livestock have largely proved them to be the work of feral dogs – domestic hunting dogs that have become untamed and aggressive, often operating in packs.

The whole concept of community conservation, where people living immediately outside conservation areas are encouraged to align themselves with the ideals of the reserve, has been particularly successful around Hluhluwe Imfolozi. This has made managing a predator population, like the wild dogs, in a non-game reserve area that much easier. What has helped enormously is that the local people are quite used to seeing predators like leopards and hyenas on their lands which, unlike the wild dogs, are dangerous to humans.

To further assist communities, project co-ordinator Brendan Whittington-Jones has come up with several practical ways to help protect livestock from predators. The most obvious is to have a herdsman accompany the animals when they are let out to graze. While this practice has to some extent fallen away in South Africa and its neighbouring states, it is still used in Kenya by the Masai living in areas adjacent to game reserves to great effect. Other suggestions include enclosing the animals in a kraal at night, tying bells around the necks of several of them to create an unfamiliar noise, leaving a transistor radio amongst the livestock to provide human sounds and keeping a guard animal such as a trained domestic dog, donkey or ostrich, which is aggressive towards predators, to ward off or attract attention to potential danger.

Armed with telemetry equipment, radios and GPS systems, researchers monitoring wild dogs in Hluhluwe Imfolozi record various data regarding pack movements, their composition, any interaction between other packs or predators and the condition of individual dogs. They take photographs of each dog for identification purposes and, if possible, collect samples of any scats to conduct DNA tests (the scat obviously needs to be linked to a particular dog). The information is collated and a database continuously updated to keep an accurate account of the current wild dog population in the park.

Carla Graaff, a researcher in Hluhluwe Imfolozi, tells of the occasion when she had been tracking the Crocodile pack for several days, with little success, when she rounded the corner on the road to Mpila Camp and spotted them resting in the shade of a marula tree. Her day was made, but it got even better when,

Opposite Carla Graaff, a researcher in Hluhluwe Imfolozi Park, maintains that each dog and therefore each pack has its own personality. This young male certainly seemed to be more introverted than the rest of the group, emphasised by his 'hang-dog' look.

a short distance away, she eyed a newly laid scat and – as luck would have it – the owner as well. Plastic bag in hand, she leapt from her vehicle and bent down to collect it, just as a small sedan pulled up next to her. The three occupants, all elderly English ladies, stared intently at her through the window and when the front passenger leaned out to say something she braced herself for the usual question 'Where are the lions?' But instead the well-modulated tones of the lady concerned congratulated her for picking up the 'doggy doo', just as they do back home!

Carla is mad about wild dogs, her light-hearted manner masking her complete dedication to their well-being. Her greatest pleasure is to be out monitoring the reserve's various packs, trained eyes flicking over the dogs and identifying each one in a tick, her research register always at hand. In the course of her work she has found that not only do the individual dogs have different colour patterns, but they have different personalities too. This gives each pack its own unique character.

The extensive DNA work being carried out in Hluhluwe Imfolozi is starting to bear fruit and a clearer picture is emerging about the dogs' genetic make-up. This will be invaluable for any metapopulation translocations in future, as it allows for the monitoring and control of inbreeding. According to Carla, this makes the reserve's dogs 'hot cakes' in the genetic stakes – but one has to wonder what they would think of the description! She maintains that a sure indicator of the presence of wild dogs in any area of the reserve is the swirling of yellow-billed kites overhead:

these birds, more than any other, are the first to swoop down to grab pieces of meat after the dogs have brought down prey.

We had made several trips to Hluhluwe Imfolozi and had still seen no sign of wild dogs, in spite of Carla and Brendan's assistance with tracking them. When the hiss and static of the telemetry equipment gave way to a steady pulse that sounded like a heart beat, we knew the dogs were near. However, on more than one occasion they were either lying up in a ravine or in a deep thicket and we couldn't reach them. By this stage we were tearing our hair out in frustration. With only a few days left in the area we were beginning to believe that wild dogs were just a myth! Then we got the breakthrough we so badly needed.

Reports came through from the game guard outpost at Cengeni Gate, in the far west of the reserve, that they'd seen six dogs near the fence in front of the staff camp the previous day. At the crack of dawn the next morning we set out from Mpila in that direction and had already covered some distance when we received an excited call from Brendan. He had just caught a glimpse of several white tails disappearing through the bush on a long sweeping bend of the White Umfolozi River, just below his house which was some three kilometres from the guard outpost.

Frantic to get there, we cursed the 40-kilometre speed limit in the reserve, did not even linger at what would have been an excellent black rhino sighting, then rounded a final corner to see Brendan's pick-up a short distance ahead. Our timing could not have been better: we'd hardly turned off the engine when the dogs materialised from the scrub in front of us, their large round ears pricked forward. Caught unawares for a moment, we looked at them and they looked at us and, as we scrambled for our cameras, they loped past and headed towards the river. Incredibly they were going to do the very thing we had hoped, for photographic reasons, they would do – cavort about in the little water that still remained in the Umfolozi riverbed, deep into the dry season.

From our vantage point on the bank we looked down at them, the long puddles reflecting the early morning light, the sounds of them splashing about were clearly audible from where we stood. For us, at that point into our own wild dog odyssey, nothing in the world could have been more magical.

In a few precious minutes we had managed to get some great shots of them as they frolicked in the water. Not only that, but

Brendan had identified a young female that had gone missing from her natal pack, as well as a male who had dispersed from Thanda Private Game Reserve several weeks before. We'd heard of the long distances dispersing dogs will travel in search of others to form a new pack and here was living proof – Thanda is more than 80 kilometres away. No wonder Brendan was so elated. What a privilege to stumble across a pack in the making. After weeks of nothing we suddenly had everything!

Opposite, above & overleaf Although wild dogs are able to live independently of water, they will drink when it is available – usually after a morning hunt or when setting out in the late afternoon. They show a reluctance to enter deep water, probably due to a fear of crocodiles.

81

IN SEARCH OF THE AFRICAN WILD DOG

This page We tagged along with the KZN Wild Dog Research team to witness them fitting a telemetry collar to a young male. To entice the pack out of the bush they dragged an impala carcass behind their truck and loudly played a recording of the dogs' amazing *hoo* call – a contact call that can travel up to four kilometres.

This page After the male was darted, section ranger Emile Smidt, Carla Graaff and her assistant Morumo Nene went through what was to us by now a familiar routine of drawing blood to screen for the presence of disease. Blood samples were also taken to add to the genetic bank of the Hluhluwe Imfolozi dogs, a project funded by the Smithsonian Institute.

Corridors of hope

Mkhuze Game Reserve is part of South Africa's first proclaimed World Heritage Site, The Greater St Lucia Wetland Park, now known as Isimangaliso Wetland Park. Mkhuze incorporates a range of habitats, from open woodland savanna to stands of thorn trees, thickets, riverine forest and an extensive area of sycamore figs. The southernmost limit of the Lebombo mountains falls within the reserve, the ancient volcanic range (geologically linked to the Drakensberg) cut by the Mkhuze River, which meanders down the eastern boundary. Central to the reserve is the magnificent Nsumu Pan, which plays host to a large number of hippos and crocodiles and hundreds – sometimes thousands – of water birds. A variety of antelope, black and white rhinos, giraffes, zebras, wildebeest and elephants inhabit the park, as do leopard, hyena and currently a single pack of wild dogs.

In spite of this idyllic setting, things do not always go the way of the dogs as they seem to have a knack of running into snares. This is a big problem in Mkhuze, and over the years many dogs have been mutilated or killed in this way. Relations with some of the surrounding communities are complex and can be strained, which makes the matter of snaring and other poaching more difficult to control. According to Paul Havemann, Conservation Manager of Mkhuze and Chairman of the KZN Wild Dog Management Group, poaching in the reserve is not all subsistence driven. Often it is carried out by or for well-organised criminals who poach specific animals to supply orders from clients, themselves involved in wider criminal activity in the region.

With improved anti-poaching measures involving pro-active law enforcement and information networks, the problem is gradually improving. So much so that poaching no longer occurs on some boundaries of the reserve. Still, the use of snares is far more common in Mkhuze than in Hluhluwe Imfolozi. It has been suggested that this could be because the Tonga people in the Mkhuze area are trappers by tradition.

Apart from the danger of snares, the wild dogs of Mkhuze run the gauntlet of competing predators and have done battle with spotted hyenas on a number of occasions – and come off second best. Fortunately for the dogs, at the moment Mkhuze has no lions. But if the fences between the park and its neighbour, the Phinda Resource Reserve (which has lions), come down as is planned, the dogs will have to deal with much more aggressive competition in the future.

The northern Zululand and Maputaland area has become strategically important for the conservation of wild dogs in South Africa through the planned establishment of so-called 'dispersal corridors' linking core populations. When dogs break away from their pack in single-sex groups, they go in search of other dogs

Above left Although the distribution of wild dogs in southern Africa overlaps uncannily with that of the blue wildebeest, they only occasionally tackle them. A dominant bull can be dangerous as he is territorial and aggressive, more than capable of inflicting damage to predators.

Isimangaliso Wetland Park, Thanda/Zululand Rhino Reserve/Somkhanda complex, Phongola Private Nature Reserve, Phongola/Nsubane Transfrontier Conservation Area, Magudu/Ithala complex and the Emakhosini/Hluhluwe Imfolozi Park complex have all been identified as suitable areas. Should the KZN Wild Dog Management Group convince the various role players in these areas to become partners in the metapopulation dispersal corridor scheme, a massive 30 000 square kilometres will become available for future wild dog packs.

A high priority is obviously to make the dispersal corridors between these regions as safe as possible for the dogs to use. To this end the initiative works in conjunction with the WWF Black Rhinoceros Range Expansion Project, as well as with other interest groups that share a similar conservation vision. Ongoing education of local communities, farmers and game ranchers to raise awareness of the wild dogs, their endangered status and their ecotourism value, has proved to be particularly encouraging in northern Zululand over the years. Exciting times lie ahead if the major provincial and private game reserves in the region make the commitment and safe dispersal corridors become a reality.

Where cattle are a man's best friend

On the hunt for traditional Zulu knowledge on the wild dogs, we picked up Zama Zwane, community liaison officer with the Endangered Wildlife Trust, and set out to visit a powerful local sangoma in the Hlabisa area, just outside Hluhluwe Imfolozi Park. Jeremiah Msezane, like other traditional healers, was called to his profession by his ancestral spirits and has special powers to communicate with them. A good-looking man, his distinctive dress added to his stature – a colourful sarong and a beaded head-dress topped by the inflated bladder of a sacrificed animal. He carried a fly whisk made from the tail of a blue wildebeest, considered to be the symbol of a sangoma's power.

With Zama as interpreter we gleaned some information about the wild dogs and, although we did not get anything in the way of Zulu folklore, we certainly learnt about their use in traditional medicine or muthi. *The Zulu language is particularly expressive and 'inkentshane' – the Zulu word for a wild dog – describes the dog's speed during a hunt and its quick movements as it eats. These days the word has a more sinister association as it is used*

in opposite sex groups to form new packs. In modern times, due to shrinking suitable protected range, populations in the region have been too small and scattered for them to do this. In recent years, however, there has been a considerable expansion and consolidation of protected areas in the north of KwaZulu-Natal, which could significantly increase the number of core populations of wild dogs.

Above Zama Zwane, community liaison officer with the Endangered Wildlife Trust, is passionate about wild dogs. He was the obvious person to assist us in our search for traditional Zulu knowledge on the dogs among those communities bordering the park.

in both rural and urban communities to describe a man who is a ruthless killer. On a happier note though, Jeremiah said his inkentshane *love charm helps a man become the 'top dog' in his lady's life – an obvious reference to the fact that in a wild dog pack it is generally only the alpha male and female that breed.* Inkentshane muthi, *made from a concoction of herbs and the powdered bones of a wild dog, is given to a man's hunting dogs to imbue them with speed.*

With that, unable to resist, he got onto the all-important subject of cattle. Close to every Zulu man's heart, cattle are an obsession and the Nguni breed in particular is highly valued. Each Nguni bull or cow has its own unique colour pattern and the praise name given to it usually describes it in terms of similar patterns found in nature: for instance, an animal that has distinct bands of black and white around its throat is given the name of mfezi *after the Mozambique spitting cobra that has the same markings. Jeremiah reeled off a list of other associations, but no mention was made of an animal being called* inkentshane – *we refrained from asking as he was in full cry and we were loath to interrupt.*

All this took some time to get through and poor Zama was quite frazzled with the translating, but it certainly brought home to us the passion a Zulu man feels for his beloved beasts. We finally got a chance to ask Jeremiah, rather timidly, what uses inkentshane muthi *has as regards Nguni cattle, particularly as individuals of both species display unique colour patterns. He considered the question gravely and then began again.... He supposed that a man who especially liked the markings of a particular wild dog could give his bull traditional medicine to replicate the same patterns in its offspring. The usual use of* inkentshane muthi *for cattle was to scatter the ashes from the burnt skin of a wild dog around the kraal, which would keep the herd together like a pack. Then he*

Above We spent a few fascinating hours with Jeremiah Msezane, a powerful local sangoma. We learnt about the use of wild dog parts in traditional medicine and were left in no doubt a Zulu man's cattle are his top dogs.

repeated the use of their scats that the young sangoma in North West province had suggested to us many months before: namely that, placed at the entrance to a cattle kraal, they would prevent the herd from dispersing when let out to graze.

With this Jeremiah paused and, with a serious look, added a proviso to his story. In using his traditional medicine, especially if it was of a particularly potent kind, he was always careful not to over prescribe as it could have the effect of passing on the negative traits of the animal, which in the case of the wild dog would be its pack mentality, its potential aggression or its need to roam.

This had been quite some session and as we took our leave even Zama, known to be able to hold his own in the talking stakes, was quiet. A quick pit stop at the local trading store to purchase cold drinks proved to be the answer, however, as one sip later and he was expounding his ideas on how to improve the lot of the wild dogs in the surrounding communities. A fun run for junior school kids, known as the Mini Wild Dog Challenge, has proved to be a huge hit and is now an annual event in Hluhluwe Imfolozi Park. Older children have also been motivated and a fire has certainly been lit under the learners of KwaGiba High School, near Mkhuze Game Reserve. They've formed a voluntary drama group with the fabulous name of Amankentshane (the plural form of inkentshane*) and at every opportunity enthusiastically perform a vibrant song and dance routine about the plight of the wild dogs.*

Although only 6 300 hectares in extent, Thanda Private Game Reserve, situated between Hluhluwe Imfolozi and Mkhuze, is big-five territory. It also has a functioning pack of wild dogs, which has proved to be part of the reserve's allure. Predator monitor Mariana Venter is passionate about 'her' dogs: the compact reserve has allowed her to observe closely the frequency and range of interactions between the wild dogs, plains game, predators and humans. The Thanda pack is by all accounts a very tourist orientated bunch as they usually den close to one of the roads, so allowing excellent viewing opportunities.

The dogs regularly enjoy the reserve's lodge facilities: they drink, bathe and play in the water feature near the reception area and use the walkways and buildings to corner prey. They have been known to turn up at weddings, for which the lodge is famous (*thanda* meaning love in Zulu), and have made at least one bride's day by hanging around long enough to be included in the official photographs of the happy couple in a game drive vehicle!

Mariana, meanwhile, believes that her protégés can do no wrong: 'They cause chaos every now and again, but most of the time it all leads up to great sightings for the guests. They killed an nyala next to the lodge one evening and the alpha male walked into the restaurant to have a look around. Imagine having a wild dog join you for dinner!'

For days in cold, wintry weather we staked out a hill, watching a hole in the rocks near the top and waiting for the Thanda pack to emerge from their den. On the third day they put in a brief appearance to sit in the sun that had broken through the clouds. By this time we were getting pretty desperate – surely they would need to hunt soon to feed their pups? Then the weather got colder, mist and rain enveloping the hill. We called it a day, planning to come back two weeks later when we would again be in the area and the weather would hopefully be more settled. It was not to be, however, as a few days later we received a call from a distraught Mariana to say five lions had found the den and killed all the pups – standard behaviour between competing predators.

We were stunned, as this was the third pack of dogs we'd been pursuing that had suffered at the hands of predators in as many weeks. Aside from the above, the Mkhuze dogs had been involved in an altercation with a group of hyenas and almost been wiped out. Meanwhile, one of the packs at Madikwe had been all but completely annihilated by a pride of lions. Was anything ever going to go the way of Lycaon pictus? *Then we heard a lone itinerant dog had travelled through community lands and – against the odds – appeared in Mkhuze to join up with what was left of that reserve's pack. Now that was a good news story!*

Overleaf Wild dogs are often mistaken for spotted hyenas (right) and at a quick glance there is a similarity between them. The early pioneers spoke of 'hyena dogs' and it was thought at one time that the wild dog was the evolutionary link between the wolf and the hyena.

Pg 92–93 The sun rising through the mist as it rolls up the valleys just outside Hilltop Camp in Hluhluwe Imfolozi Park can be quite stunning, especially during the winter months. Steep forested hills with grassy slopes and riverine woodland are a feature of the Hluhluwe section of the park.

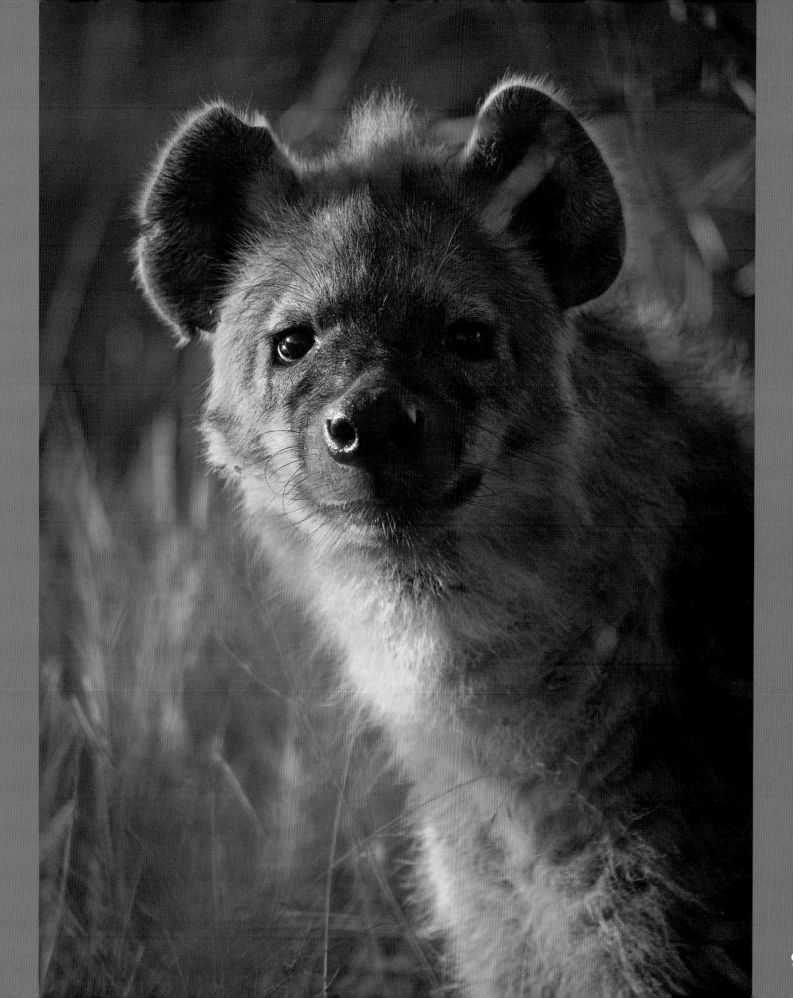

AmaNkentshane ~ Zulu

Left The anti-poaching unit in Hluhluwe Imfolozi is a formidable team. While their attention is currently directed at an upsurge in rhino poaching, they do not overlook anything untoward during their patrols. Snaring is not a major problem in the park, which is good news for the dogs as they are vulnerable to snares.

IN SEARCH OF THE AFRICAN WILD DOG

This spread Few of the rivers in either Hluhluwe Imfolozi or Mkhuze game reserves contain permanent water. As the game congregates around the remaining pools, the tracks they leave behind tell their own story, like those made by a wild dog in the coarse wet sand of the Umfolozi River.

IN SEARCH OF THE AFRICAN WILD DOG

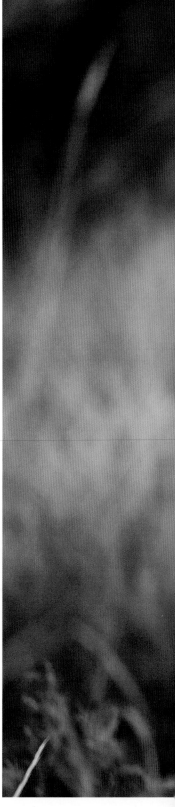

This spread The reintroduction of wild dogs to Hluhluwe Imfolozi during the early 1980s has proved particularly successful. Previously exterminated as vermin by farmers and game rangers alike, they had not been seen in the area since 1933. Today there are some 80 individuals in 10 packs.

AMANKENTSHANE ~ ZULU

What is the Future of Wild Dogs in KwaZulu-Natal?

Paul Havemann, chairman of the KZN Wild Dog Management Group comments:

The current status of wild dogs makes the KwaZulu-Natal population a very important one in South Africa as it holds between 20 and 25 per cent of the estimated national population. The challenge of this population is that 95 per cent of these animals occur in Hluhluwe Imfolozi Park.

The region has the potential of hosting a sub-population that could well be as important (or even more so) than Kruger's, provided additional core populations are established in all the major provincial parks in northern KwaZulu-Natal (Ithala, Phongola, Tembe) as well as in key private game reserves (Phinda, Somkhanda, Zululand Rhino Reserve).

What is crucial for the population to thrive is the creation of more opportunities for opposite single-sex dispersing groups to meet up with each other, and for opportunities (although rare) for dispersers to find other packs. The existing population is already defining its dispersal corridors, conservationists now need to deal with the social, economic and political issues to safeguard the way for itinerant wild dogs between new core packs.

Where to See Wild Dogs in KwaZulu-Natal

Ezemvelo KZN Wildlife
(Hluhluwe Imfolozi Park, Mkhuze Game Reserve)
  Tel: +27(0)33-845-1000 / Fax: +27(0)33-845-1001
  E-mail: webmail@kznwildlife.com
  Website: www.kznwildlife.com

Thanda Private Game Reserve
  Tel: +27(0)11-469-5082 / Fax: +27(0)11-469 5086
  E-mail: reservations@thanda.co.za
  Website: www.thanda.co.za

FOUR

Matlhalerwa

(Venda/Tswana)

LIMPOPO VALLEY

Left On their way to roost, a flock of guineafowl scurry along the banks of the Limpopo River at sunset in the Northern Tuli Game Reserve. Their shrill cries cut through the hot air and startle every living thing about.

It had been almost three years since we had crossed the Limpopo River from South Africa into Botswana to visit the Northern Tuli Game Reserve and both of us were feeling pretty emotional. We had spent almost 11 months there photographing our book Tuli – Land of Giants *and, as we swallowed hard to dislodge the lumps in our throats, it was fitting to think we had a pack of wild dogs to thank for our return. Special animals in a special place, what more could we have asked for?*

In the legendary Limpopo valley, where South Africa, Botswana and Zimbabwe meet, is the proposed location of the Limpopo/Shashe Transfrontier Park, which will straddle all three countries and cover some 5 000 square kilometres. The park will include spectacular scenery, a great diversity of wild animals, more than 350 recorded bird species and some 26 different plants that are on the Red Data List.

Rich in both history and prehistory, the area is part of an ancient landscape which, for tens of thousands of years, southern Africa's original inhabitants – the Bushmen – called home. Their nomadic hunter-gatherer lifestyle remained unchallenged for centuries until around 2 000 years ago when Bantu people migrated into the region from east and west Africa. Their arrival heralded the dawn of the Iron Age in southern Africa. There are several archaeologically important Iron Age sites in the valley, the Mapungubwe complex in South Africa being of particular significance, especially since its proclamation as a World Heritage Site in July 2003. Mmamagwa in Northern Tuli was a satellite city of Mapungubwe.

The arrival of black settlers during the first millennia AD took place in a series of waves and brought dramatic changes to the region. Through the centuries their numbers grew and their communities prospered. They cultivated grain and kept cattle and were metal workers and potters. Most importantly they were traders – part of a vibrant trade network controlled by Arab merchants who sailed the Indian Ocean between east Africa, Arabia and India. Imported glass beads, silk cloth and ceramics were exchanged for ivory, gold and rhino horn. By the time of the establishment of the Mapungubwe Kingdom in 1220, a highly sophisticated society had evolved in the valley. Mapungubwe

Opposite Obvious aggression between members of a pack is extremely rare and indeed adults usually vie to be the most submissive. This behaviour involves licking the face or tugging at the lips of the other, all the time twittering intensely. This harmony sets a wild dog pack apart from other predators, like lions and hyenas.

flourished until around 1290 when cultural, political and economic power shifted to Great Zimbabwe.

More recently, in the second half of the 1800s, the area witnessed the rivalry between two powerful chiefs, Lobengula of the Matabele and Khama the Great of the Bangwato, which had a ripple effect on communities throughout the region. It also felt the ruthless ambition of Cecil John Rhodes and President Paul Kruger and several of the opening conflicts of the Anglo Boer War. Rhodes often traversed the region, obsessed with empire building and his vision of a Cape-to-Cairo railway line. The Zeederberg coach route from Pretoria in the Transvaal Republic to Bulawayo in Matabeleland came through this part of the Limpopo valley and transport riders, their wagons loaded with supplies, mail and passengers for further north, were a regular sight. After the declaration of war by President Kruger in October 1899, Boer and British forces traded insults and gunfire across the Limpopo River for a period of some three months before both sides withdrew to defend other fronts.

An old vision rekindled

The idea of a Transfrontier Park spanning the confluence of the Limpopo and Shashe rivers is not a new one. It was first envisaged as far back as 1922 by Prime Minister General Jan Smuts when he established the Dongola Botanical Reserve in the region. In 1947 the reserve was renamed the Dongola National Park and it was at this time that the possibility of joining it with similar conservation areas in neighbouring Botswana and Zimbabwe was formally considered. But when the Afrikaner Nationalist Party grabbed power the following year, one of many acts to erase the symbols of its political foes was to de-proclaim the reserve.

Fortunately, in 1967 the Vhembe National Park was established to protect the archaeological significance of Mapungubwe (as part of the Nationalists' 'separate but equal' polices, in this case for a future Venda homeland). Over the years the land it

encompassed was significantly extended and in 2004 it was officially revived as the Mapungubwe National Park. This was a big boost to conservation in the region and provided momentum to the cross-border game reserve plan.

After decades of local extinction a number of game species have been re-introduced into the Limpopo Valley: these include roan antelope, tsessebe, black and white rhino and wild dog. The first wild dogs were relocated to the Venetia Limpopo Nature Reserve in South Africa in 2002 and to the Northern Tuli Game Reserve in Botswana in 2008. Both reserves are large enough to host their own wild dog populations and packs have indeed re-established themselves in both reserves, becoming part of a landscape from which they had been noticeably absent. Together these reserves are now core areas of the new Limpopo/Shashe Transfrontier Park, which will provide a vast range for future packs.

Venetia reserve was established in the late 1980s by the De Beers Consolidated Mines diamond company. It is situated some 10 kilometres south of the Limpopo where that river describes its most northerly arc before reaching the Kruger National Park. The reserve is some 40 000 hectares in extent and shares the same rugged terrain, and extensive mopane bushveld as its neighbour, the fairly newly established Mapungubwe National Park. Venetia is home to numerous antelope species, zebra, giraffe and four of the big five – lion, elephant, rhino and leopard – but in recent years it is perhaps its wild dog population that has attracted most visitors to the reserve.

Having assisted in successfully establishing a functioning pack of wild dogs in the reserve, the Venetia Wild Dog Project – an

Above Lobengula became King of the Matabele in 1870. He was reportedly a huge man, over six feet tall and weighing more than 300 lbs (136 kg). He considered the Limpopo Valley to be his royal hunting ground and regularly received visits from early white hunters and traders looking for concessions in the area.

initiative of the Endangered Wildlife Trust and Oxford University's Wildlife Conservation Research Unit – has embarked on a programme to give the dogs a better public image. It's no easy task as there are numerous hunting operations in the area as well as a lucrative market in the sale of game. Landowners are naturally suspicious of any carnivore that could kill their stock and affect their source of income. The project hopes to highlight the potential economic benefits of ecotourism generated by the presence of the wild dogs, which could go some way to counteracting the costs of their consumption of antelope. You cannot confine wild dogs even in a large reserve, so they'll need all the good press they can get in the Limpopo valley now and in the future. Like other packs of wild dogs in the South African metapopulation, several of the adult dogs in Venetia have been fitted with radio collars. The pack is monitored by researchers almost daily to collect ecological data regarding their behaviour, hunting and dispersal, with particular reference to packs in small reserves. The dogs in Venetia, whose number has varied significantly over the years, have been the subject of a five-year PhD research programme by Harriet Davies-Mostert of the Endangered Wildlife Trust.

Long interested in threatened species, more specifically the creation of metapopulations of large carnivores as a means for their survival, Harriet specialised in wild dogs almost by chance. She and a colleague were involved in research in the Matusadona National Park on the shore of Lake Kariba in Zimbabwe, both of them keen to work with cheetah. But a choice had to be made: the friend got the cheetahs, she got the wild dogs and she's never looked back. Now head of the Carnivore Conservation Group and chairperson of WAG – The Wild Dog Advisory Group, Harriet is something of a mover in wild dog circles.

A new kind of boundary

Just across the way from Venetia, on the opposite bank of the Limpopo in Botswana, is the 71 000-hectare Northern Tuli Game Reserve. It is owned by an assortment of commercial lodges, syndicates and individuals who, in the early 1960s, banded together to form one of the world's largest private game reserves. The area had previously been farmed for centuries, the land cultivated and cattle and goats kept. Today it is home to a sizeable elephant population, Africa's big cats – lion, leopard and cheetah – herds

of giraffe, zebra and wildebeest and a range of antelope from the giant eland to the tiny steenbok.

In April 2008, having spent six months bonding in a boma, a pack of 18 wild dogs was released with great fanfare into this land of huge vistas and big skies. In spite of the suitable terrain, a low density of other carnivores and large herds of impala, any wild dogs previously seen in the reserve have been individuals or small groups that had moved through and chosen not to stay. With luck a resident community of dogs will now encourage any dispersing groups from other packs to remain in the area by providing possible mating opportunities and in this way establish a viable population of wild dogs in the region.

Above The Venetia Limpopo Nature Reserve is scenically dramatic and shares the same majestic terrain as its more famous neighbour – the Mapungubwe National Park. A pack of wild dogs was reintroduced into Venetia in 2002.

Matlhalerwa ~ Venda/Tswana

Left The wild dogs in the Venetia Limpopo Nature Reserve have proved to be one of its biggest tourist attractions. While the dogs endure a high incidence of dangerous confrontations with lions, it is perhaps the formidable fence that separates the dogs from the adjoining diamond mine that epitomises their tenuous existence.

We were lucky enough to be present for the release and the emotion of the occasion affected everybody present – especially researcher Craig Jackson from The Mammal Research Institute of the University of Pretoria, who had nurtured this pack through its long stay in the boma. As the gates opened and the dogs followed his pick-up out of the enclosure, lured by the freshly killed impala he had attached to the back, a cheer went up from the small group of spectators gathered to see the pack's first steps to freedom. There was no sentiment attached to any of this for the dogs, however, as they were intent only on getting their teeth into the impala carcass as quickly as possible.

This was how they had received their food for the previous six months in the boma – off the back of a pick-up – and it was how they expected things to continue. For several days after their release they hung around the main road through the game reserve, which is also the route to the towns of Bobanong and Mothabaneng from the Pont Drift border post and therefore quite busy. This caused Craig huge concern as wild dogs are fatefully oblivious of fast-moving vehicles. To make matters worse, they ran after every truck that passed in the hope it would offload their next meal.

By the third day they still showed no sign of setting off to hunt and Craig was getting frantic. Then a warthog family inadvertently trotted right into the middle of the pack resting up in the shade at the edge of a dry riverbed. This triggered an immediate response in the dogs, their natural instinct to hunt taking over and in a blink of an eye they brought down one of the piglets and devoured it. It was no more than a snack for a few of the dogs but it had the desired effect of turning their attention away from vehicles as a food source to the abundant prey in the reserve.

Left **The** most successful reintroductions of wild dogs have been those where the dogs were first bonded in a boma and released only after they had formed a socially integrated and cohesive pack. The 18 dogs reintroduced into the Northern Tuli Game Reserve in April 2008 spent six months penned in before their release.

This page As release day dawned, researcher Craig Jackson and community liaison officer Rex Masupe off-loaded a dead impala to drag behind their pick-up and lure the pack out of the boma. When the dogs followed the carcass out the gates, there was not a dry eye amongst the spectators who had gathered to witness the event.

Overleaf The speed at which the dogs operate is staggering to any human. The camera data recorded while filming this incident shows it took just two seconds for them to catch a warthog piglet and 20 seconds to devour it.

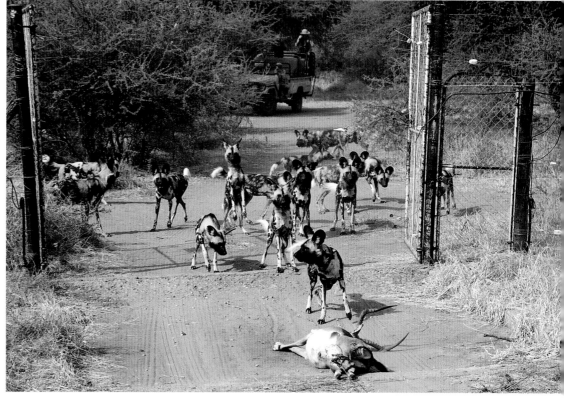

In Search of the African Wild Dog

Matlhalerwa ~ Venda/Tswana

Matlhalerwa ~ Venda/Tswana

We returned to Tuli some two months later, in response to Craig's call that the pack had denned in a thicket of fever-berry, or croton, trees on the banks of the Limpopo. By then the dogs had become ace hunters and were regularly bringing down impalas and larger antelope, including quite a sizeable eland calf that had strayed from the herd. We timed our visits to the den site for around 17h00. Within about 30 minutes the alpha female would rouse the rest of the pack with a rallying call of twitters and squeals and they would head out to hunt. The dogs had perfected a technique whereby they would single out their prey and then run it up against the fences around the Merry Hill workshops at the Pont Drift border post, only about a kilometre from their den and within easy reach to rush back to regurgitate meat for the pups. On one occasion they chased three different impalas into the bush adjacent to the workshop, instead of trying their tested technique, and each time they met with no success.

We raced around in Craig's vehicle trying to follow them, but they were way too fast and we kept losing them in the scrub. Then, for once, we managed to second-guess their next move and sped towards the staff soccer field nearby. We had hardly reached it when we heard the familiar frenzied twittering that signals a kill and saw that the dogs had brought down a sub-adult kudu bull against the fence of their old home, the boma.

It wasn't easy to watch or to take photographs as the kudu seemed to take a long time to die, the dogs tearing at its belly and innards. The fated animal made lunging movements with its long horns towards the dogs as they fed, trying to fend them off and a number of them had to leap away to dodge the thrusts. But to no avail. As twilight gave way to darkness the kudu succumbed to its wounds and loss of blood and we were only too pleased to leave the pack to its meal, though the gory scene stayed with us for a long time afterwards.

Left & overleaf Wild dogs have a reputation of being ruthless and indiscriminate killers, no doubt brought on by their method of killing. Once they have exhausted their prey they disembowel it, tearing at its vital organs while it is still standing, which is not pretty to see. They sometimes use man-made structures to assist them during a hunt, as in this case in the Northern Tuli Game Reserve when they cornered a young kudu bull against a fence.

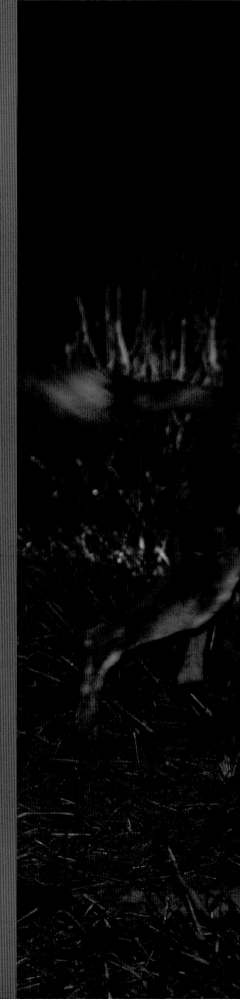

The release of dogs into the Northern Tuli Game Reserve kick-started an experiment with a new and fascinating concept, the so-called 'bio-boundary'. For months before their release Craig had placed hundreds of scent markings and scats, collected from other wild dogs, along the perimeter of the reserve at regular intervals to simulate the territorial markings of another wild dog pack. In theory the released dogs would pick up on the chemical stimuli or 'messages' in the scats and not stray beyond the confines of the reserve into community lands where they could be killed. Brainchild of Dr Tico McNutt of the Botswana Predator Conservation Programme, the bio-boundary holds promise for future translocations and could become a powerful tool in limiting potential conflict between wildlife and affected communities.

Above Craig Jackson is pioneering the use of 'bio-boundaries'. Hundreds of scats collected from wild dogs elsewhere have been placed along the western perimeter of the Northern Tuli Game Reserve to simulate the territorial markings of another pack. The 'boundary' appears to have successfully dissuaded the Tuli dogs from crossing into community lands.

In the course of his research Craig clocked up thousands of kilometres as he monitored the Tuli dogs' reaction to their bio-boundary. He found the linear pattern of the scats and scent markings was not quite sufficient as once the dogs had crossed this line (as happened on a few occasions), there were no further clues to indicate to them that another pack occupied the area beyond. It became obvious that additional scats, randomly placed behind the original line, would be needed and this put quite a strain on Craig's resources. His uniquely personal SOS – send out scats – was regularly relayed to Dr McNutt who never failed to oblige with a consignment of deep frozen, foil wrapped samples!

Craig is now much more satisfied with the new system of distributing scats:

After leaving the den, the wild dogs are roaming the reserve again and the last few weeks have seen them moving in a westerly direction along the Limpopo River. This was a favourite area for them prior to denning, when they spent a good deal of time to the west of the veterinary fence; but not ideal as there are several cattle and goat posts in the vicinity. We had expected the dogs would move back to this area, but this time round we staggered the marks so that we effectively had a zone or belt of scent marks, and not merely a single line. After moving west along the Limpopo, the dogs encountered the scent marks and gradually moved up and away from the river, following the bio-boundary very closely. They then moved back east, which is exactly the type of response we would like to see. This, in combination with behavioural observations, clearly indicates to us that the wild dogs are responding to the artificial scent-marked territory we've created. Hopefully we will be able to document similar responses when next the dogs encounter the bio-boundary.

Opposite Wild dogs are highly specialised hunters. Teamwork, speed and stamina make them formidable predators, but it is their ability to stay focused that accounts for their hunting success. Once the leader has selected a quarry, he or she locks on to it; when the rest of the pack joins in they focus on the same animal, oblivious to any other.

Goat dogs in Botswana

Kelebileone Oboletse is a ranger at Tuli Safari Lodge with a wealth of both conventional and traditional knowledge of the bush. He gained the latter from his grandmother who is a sangoma, or traditional healer, and who he says is responsible for his almost impossible sounding name that means 'looking towards God'. (She had apparently longed for a son but had only managed to produce daughters, so his birth was the answer to her prayers.) For some reason, much to our relief, he has been called Moscow ever since he first started work as a tracker and game ranger many years ago – perhaps by some Russian guests!

When we asked him about the wild dogs over breakfast one morning, Moscow said the Tswana word for them in the north and central part of Botswana, and therefore in the Limpopo valley and Tuli area, is matlhalerwa, as opposed to the word makanyane used further south. Both words allude to the same thing, however – the dogs' agility during a hunt. He admitted that his community near the Okavango Delta disliked them and poisoned or shot them on sight because they believed the dogs would kill their livestock and threaten their livelihood. To our horror he mentioned that some people actually set fire to dens to kill the pups inside.

Seeing our complete dismay at his last comment, Moscow moved on to add that the community of Loensa le Moriti, whose lands fall within the reserve, had devised an ingenious method of protecting their livestock from predators, which did not involve killing them. This was through the use of so-called 'goat dogs'. At birth, a puppy is removed from its natural mother and given to a female goat to suckle from and imprint on. It stays with its goat family throughout its life, going out into the veld during the day and returning to their enclosure at night. But because of its canine genes, it barks to ward off unwelcome visitors, alerting the community to the presence of a predator looking to grab any goats. This was a great story, a truly local version of the Anatolian guard dog initiative used successfully by more affluent farmers and game ranchers in Namibia and South Africa.

As Moscow plied us with eggs and bacon, he proudly announced that his grandmother used a variety of fly whisks (a symbol of a sangoma's powers) in the course of her work. Not only did she use the tail of a blue wildebeest, which is the most common, but she also had one made from a horse's tail and another from the tail of a wild dog. Her choice of whisk depended on her mood

Left The strength of the species lies in the tight social structure of a pack, which enables them to nurture their young and hunt effectively. A lone wild dog has little chance against the harsh realities of nature.

Opposite A wild dog has a substantial lower jaw with well developed muscles, which give it its hugely powerful bite. The cheek teeth are serrated and are used for tearing meat off its prey.

IN SEARCH OF THE AFRICAN WILD DOG

Above For communities bordering many of southern Africa's game reserves the best way to protect their livestock from predators is for somebody to watch over the animals while they are out grazing. But like young Kutlwano Pilane (above with stick), most would-be herd boys now attend school from early morning to late afternoon.

and on the occasion, but she used them in divinations with the ancestors. After they had been swished through water infused with special herbs, she would flick droplets around a homestead or over a person to dispel evil spirits or give a blessing. She also prescribed traditional medicine made from the powdered bones, teeth and nails of a wild dog to instil courage and speed in a man's hunting dogs, or to give his cattle an air of strength to make them impervious to predators.

The Northern Tuli Game Reserve has followed the example of Hluhluwe Imfolozi in KwaZulu-Natal: embarking on a community awareness campaign to bring about respect for the wild dogs and so create a safe environment for them, should they extend their home ranges into areas adjacent to the reserve. Like in KwaZulu-Natal, these communities are quite used to seeing predators such as hyenas and leopards on their lands. The programme therefore emphasises this acceptance needs only to be extended to the wild dogs. Better management of livestock is stressed, as is the fact that the dogs actually prefer to eat antelope as opposed to goats or sheep. The ecotourism benefit of having the species in the reserve, resulting in financial spin-off to the surrounding communities, is also highlighted. An increase in tourists also means more employment for local people as staff for the lodges are drawn preferentially from these areas.

Community liaison officer Rex Masupe comes from the neighbourhood and this helps immeasurably when dealing with any problems arising from incidents with predators that have strayed outside the reserve. Like his father, Rex is a long-standing member of the reserve's anti-poaching unit, which gives him added credibility when addressing any concerns or queries.

Rex and his team regularly pick up snares that have been set in the riverbed and which pose a threat to all the animals in the reserve as there is no fence along this boundary. The Endangered Wildlife Trust works in conjunction with the Northern Tuli Game Reserve and the Venetia Limpopo Nature Reserve on the anti-poaching front. They also co-operate in 'farm extension work' to promote the concept of a transfrontier park, helping to reduce human–predator conflict among neighbouring South African farmers – particularly those who farm along the Limpopo. It is hoped that all these measures, when taken at the flood, will turn the tide in establishing a viable wild dog population in the region.

Above Rex Masupe, community liaison officer with the Tuli Wild Dog Project, works tirelessly to reinforce a positive image of wild dogs and so create a safe environment for them. He comes from the region, which gives him credibility in his dealings with community members when resolving any human–predator conflict.

Overleaf As dawn breaks on a chilly July morning, a herd of eland crosses the Limpopo River. Africa's largest antelope, an adult male can weigh an incredible 940 kilograms. That puts it beyond reach as wild dog prey, but dogs are known to bring down eland calves.

In search of the African Wild Dog

This spread Wild dogs have an incredibly strong reproductive drive and while it is usually only the alpha pair that breed, the rest of the pack shares in the collective parenting of the pups. Care for the young continues even after they are big enough to join in the hunt.

Matlhalerwa ~ Venda/Tswana

Right The dogs concentrate on whatever prey is most abundant in their area – big or small. In southern Africa that means antelope seldom larger than waterbuck; however, in east Africa packs hunt wildebeest and zebra far more frequently.

What is the Future for Wild Dogs in the Limpopo Valley?

Harriet Davies-Mostert, chairperson of WAG comments:

Northern Limpopo province is one of the few areas in South Africa where wild dogs roam free outside of protected areas. Although the status of these packs is tenuous (they occur at very low densities and are vulnerable to persecution from farmers), the fact that they occur here at all is testament to the suitability of the terrain for their survival. The two wild dog re-introductions into the Limpopo valley – one into the De Beers Venetia Limpopo reserve and the other into Northern Tuli – are likely to improve their outlook.

By establishing resident packs in the area, these reintroductions will provide opportunities for natural pack formation with any dispersing groups that happen to pass through. As the Limpopo/Shashe Transfrontier Conservation Area becomes a reality, and fences come down between South Africa, Botswana and Zimbabwe, the reserve network could grow to encompass more than 5000 square kilometres. This would support a significant number of wild dog packs and also provide an important corridor for wild dogs moving between Botswana, Zimbabwe and the northern Kruger complex.

Where to See Wild Dogs in the Limpopo Valley

Mashatu Game Reserve

Tel: +27(0)11-442-2267 / Fax: +27(0)11-442-2318

E-mail: reservations@malamala.com

Website: www.mashatu.com

Mopane Bush Lodge

Tel: +27(0)83-679-8884 / Fax +27(0)15-534-7906

E-mail: info@mopanebushlodge.co.za

Website: www.mopanebushlodge.co.za

Nitani Private Game Reserve

Tel: +27(0)31-764-2346 / Fax: +27(0)31-764-2179

E-mail: reservations@nitani.co.za

Website: www.nitani.co.za

South African National Parks (Mapungubwe National Park)

Tel: +27(0)12-428-9111 / Fax: +27(0)12-343-0905

E-mail: reservations@sanparks.org

Website: www.sanparks.org

Tuli Safari Lodge

Tel: +267-264-5303/43 / Fax: +267-264-5344

E-mail: info@tulilodge.com

Website: www.tulilodge.com

Venetia Limpopo Nature Reserve

Tel: +27(0)15-534-2986/575-2651

E-mail: reserve@debeersgroup.com.

FIVE

Mahlolwa

(Shangaan)

GREATER KRUGER NATIONAL PARK

Left The high-level bridge across the Sabie River in the Kruger National Park is a great spot to be at sunrise or sunset. The river is the largest in the park and its perennial waters are home to numerous hippos and crocodiles.

It's a truism that the Kruger National Park, sprawled across some 20 000 square kilometres, is larger than some countries, which is quite daunting really if you are on the lookout for the world's most restless traveller – the African wild dog! There are only between 180 and 450 dogs in Kruger (their numbers fluctuate hugely) and this small group has the status of being the only genetically viable population in South Africa. With these odds, it makes coming across any of them very lucky indeed.

The Kruger National Park is one of South Africa's foremost icons and perhaps its greatest tourist attraction. Established in 1898 as the Sabi Game Reserve, it was expanded over the years. The park finally achieved national status in 1926 when it was given its present name in honour of President Paul Kruger of the old Transvaal Republic – even though he actually opposed its proclamation in the Volksraad.

The park lies in the north-eastern corner of the country, beyond the Great Escarpment, in an area most suitably called the lowveld: a region of hot bushveld, fever trees, malaria, crocodiles and big game. It bumps up against Mozambique in the east and Zimbabwe in the north and is home to a huge diversity of plant, bird and animal life – said to be more diverse than in any other conservation area in Africa.

The region has been inhabited by humans for millennia and several significant historical and archaeological sites are found within the park. South Africa's first known inhabitants, the Bushmen, used a number of the caves and rocky shelters and left behind a legacy of Stone Age tools and artefacts, as well as numerous rock paintings, like those at Hippo Pools near Crocodile Bridge in the south of the park (unfortunately all but obliterated by exceptional flooding in 2005). Iron Age people also lived here. A walled settlement on the hilltop at Thulamela near Pafuri in the far north provides insight into the people who settled in the vicinity between the 15th and 17th centuries.

Thulamela is the Venda word for 'place of birth' and myths and legends abound about the area. Gold beads, spindle whorls, ornamental cowry shells and various metal and ivory rings unearthed

---

Opposite The wild dog belongs to the genus *Lycaon*, which formed a new branch on the canid family tree some three million years ago. It is the sole survivor of this evolutionary branch and, because of its genetic difference, is unable to interbreed with any of its canid relatives, including the domestic dog.

on site provide proof that the community was part of the dynamic trade network controlled by Arab merchants off the east coast of Africa during this period – as discussed in the previous chapter. The Venda were (and still are) a very mystical people who lived in touch with the spirit of the land. The ancients associated different stars with different animals, like a giraffe, a wildebeest or a wild dog. In each case when they could see the star they related it to certain behaviour of that particular animal and it gave harmony and meaning to their universe. The shape of the main dry-stone wall structure at Thulamela, surrounded by a natural encirclement of baobab trees, resembles the totemic animal of the ruling dynasty – a crocodile.

By the mid-1800s the abundant wildlife found on the vast plains of the highveld and later the lowveld had become the target of wholesale slaughter by hunters, traders and trophy seekers.

Before the discovery of gold on the Witwatersrand, game harvesting was the economic mainstay of the Boer Transvaal Republic. Even prior to the establishment of the Sabi Game Reserve, the fabled great herds had pretty much disappeared and what animals did remain were worryingly low in number.

The great mover and shaker

Colonel James Stevenson-Hamilton, on assuming his post as warden in 1902, was stunned by what he saw: the area had been almost completely denuded of big game during the Anglo Boer War. The appointment of the colonel, a Scottish cavalry officer who had fought in that war, proved to be an inspired choice. For more than 44 years he toiled tirelessly in steering the park towards the wildlife sanctuary it is today. He had a reputation for not suffering fools gladly and his ruthless approach in stamping out poaching earned him the nickname of *Skukuza* from the local Shangaan people. The name, which means 'the one who turned things upside down', was perhaps kinder than deserved as some of the first 'poachers' Stevenson-Hamilton evicted were a small Shangaan clan who had settled in the area near the present-day Tshokwane picnic site, long before the establishment of the park. (Tshokwane was the name of their chief.) The park's largest camp, on the Sabie River, is called Skukuza after Stevenson-Hamilton.

His various accounts of his experiences as game warden make fascinating reading. Not only do they reflect a deep knowledge of the African bush but they also provide a record, however subjective, of the state of the different species of game in the Sabi Game Reserve/Kruger National Park during his tenure. Some of his observations on wild dogs highlight not only their situation but also that of their principle prey during the early 1900s:

> When wild dogs were numerous in the Sabi Game Reserve, it seemed to me those found in the neighbourhood of the Olifants River and northwards tended to be smaller and lighter, while those south of the Sabi River were larger and darker. The natives [sic] used to say that there were two species of wild dogs in the reserve, one larger than the other.
>
> The pack which formerly hunted the country between the Sabi and Crocodile Rivers in the Sabi Game Reserve, and which ranged over an area of some fifteen hundred square miles, well stocked with impala and reedbuck, consisted, before it had been artificially reduced in numbers, of between sixty and seventy individuals. A second pack, having a very wide range from the Sabi on the south to the Olifants River on the north … was never quite so numerous….

Above Even though President Paul Kruger of the Transvaal Republic personally tried to block the proclamation of a game reserve in the lowveld, it later proved politically expedient to use his name in order to drum up support for the creation of South Africa's first national park.

There is no other predatory animal in Africa responsible for so much disturbance of game as the hunting dog, and in proportion to his numbers, there is none which deals out more wholesale destruction. A pack, descending suddenly upon a district, scatters the animals far and wide, and it is not difficult to discern from the restless and uneasy manner of the impala and other antelopes, as well as from the fact of the herds being split up and scattered into small parties, that these bush pirates are on the warpath.

During the lambing-time of the impala and reedbuck, the young of these species seem, together with the parturient ewes, to form the principal prey of *Lycaon pictus*, and the number of recently born animals destroyed is enormous.

I remember following a pack of about thirty, which for two days about the end of November had been playing havoc among the young impala near Sabi Bridge. I counted at least a dozen kills within a radius of less than a mile, all those of small lambs. An impala of about a week old forming one meal for a single dog, and there being but little difficulty in running it down, the harm that is done to the increase of the herds in this way may be easily realised.

I crept up to three dogs which were busily employed in eating a young buck, and was struck by the frenzied haste which they showed in rending it in pieces and bolting it in huge pieces. One instance will serve to indicate the rapidity with which these animals dispose of their food. Two native rangers [sic], stationed with one of our pickets, heard a duiker cry out close by.

Pausing only to seize their spears from the hut, they ran to the spot, hoping of course to secure some meat for themselves; but although between the time of the first alarm and their reaching the place barely three minutes could have elapsed, nothing remained on the ground except a little blood and some pieces of skin, while three or four hunting dogs could be seen decamping through the bush.

Right Baboons form tight social groups and when the troop is threatened the larger males join ranks to form a formidable front. Few predators will stand their ground against such a ferocious display, but will snatch any stragglers hanging about. Their greatest foe, the leopard, usually tries its luck after dark.

Wild dog hide-and-seek

We searched the bush in the south of Kruger for weeks on end, the area around Pretoriuskop, Skukuza and Berg-en-Dal camps being the location of sporadic sightings of wild dogs at the time. But try as we might we had no luck. 'They were here last week', 'they moved off a few days ago' and 'you just missed them' were phrases we came to dread and in the heat and dust our sense of humour began to wane. Then – as will happen in the bush – we hit the jackpot! Reports came through that a pack of wild dogs had denned in the Manyeleti, a game reserve much further north, which bordered on the Kruger's unfenced western boundary. If there was a den it meant sightings were a dead cert. So, after seeing out our stay in Skukuza, we headed up two days later to Tintswalo Safari Lodge and soon found ourselves sitting a few metres from the den with head ranger Adrian Bantich.

The den, in an old termite mound, was ominously quiet. As an unseasonably cold wind sprang up and whipped through us, we had the first stirrings of doubt as to whether the dogs were actually still there (though perhaps it was just the wind that was keeping them below ground). As the sun set and evening moved in we were forced to return to camp with no sighting of the dogs.

Above Mixed wooded grassland areas in the Kruger are a favourite haunt of the wild dogs. They provide good cover after a kill as well as places to hide from other predators. Hunting conditions are improved during dry periods when ground cover is sparse and the game more visible.

Opposite While the enmity between lions and wild dogs is not as well documented as that between lions and hyenas, wild dog packs are unlikely to challenge lions the way hyenas will at a kill, and will always come off badly in any confrontation.

To us this seemed almost unbelievable as the five adults and their eleven pups had been seen outside the den for many weeks and, more importantly, on the afternoon before our arrival. When the following day still brought no sign of them, in spite of extensive searching, it became horribly obvious that the pack had moved the pups and there was every chance we were going to miss out on what had earlier seemed a certainty.

But where was the new den and how far away was it? With the offspring being only about two months old and not yet weaned, we reckoned they would stay relatively close to their original site. The area was littered with old termite mounds, similar to the one they had first chosen. The next day dawned, our last at Tintswalo, and as we set off we hoped like mad that our strategising the night before would work. The idea was to head back to the original den, make absolutely sure that there were no fresh signs of the dogs there and then fan out. We'd move from termite mound to termite mound, using binoculars to scan the vicinity, while keeping our ears open for any wild dog sounds.

We headed north and Adrian went east.

We were working an area about 100 metres square and had climbed several mounds when we heard the faint sound of a branch snapping. Three elephant bulls off to our right! A few minutes earlier we had noticed several white rhinos in the distance. After gesticulating frantically to Adrian to warn him about their presence we climbed yet another mound, pleased that we were downwind of both species. And then Adrian, who was keeping a sharp eye on the approaching elephants, caught a glimpse of the dogs outside their new den, some 150 metres away in the direction the elephants were headed. We ducked back to the vehicle and carefully approached, anxious not to disturb the pack. The relief at finding them was enormous: after weeks in the Kruger and with time running out, this was our last chance to get any photographs of wild dogs in the area.

Then, unbelievably, one of the dogs got up and darted towards the elephants in what looked like an attempt to distract them, as they were by now very close to the den. Outraged by this effrontery, the bulls postured and threatened and then calmed down, the two younger ones moving behind us where they continued to browse. Then suddenly, for no apparent reason, the third bull stormed down on the dogs, scattering them in all directions, the pups fortunately hidden safely away below ground. He stopped just metres away from the entrance to the den, abruptly losing interest but having demonstrated his dominance, then walked off to join his companions. What a morning – from desperation to elation in just a few minutes; but that's wildlife photography!

Adrian glowed with delight, Manyeleti's 'north pack' had shown their mettle once again! By all accounts the pack is quite fearless, their interaction with the elephants not all that incredible. A few months earlier Adrian had witnessed the dogs nipping the ankles of a buffalo calf, in the midst of a herd of some 40 buffaloes. The bleats of the calf and the distress calls from the mother alerted a group of bulls that were grazing nearby and they turned as one on the dogs, bellowing and thundering towards them.

The pack escaped from the melee unhurt and obviously undaunted, but their behaviour in both incidents becomes all the more startling if one considers that an adult dog weighs a mere 30 kilograms, an adult buffalo some 750 kilograms and an elephant bull a whopping five tons!

Two dogs of Africa

Mahlolwa is the Shangaan name for a wild dog and means 'taboo', in the sense that the dogs' method of killing is considered unpleasant and not something that one wants to witness. Shangaan sangomas 'throw the bones' to consult the ancestors in order to help their patients solve physical, emotional or spiritual problems. The bones or charms they use are called *tinhlolo*, which is apparently a derivative of the word *mahlolwa*. A concoction made from the powdered teeth of a wild dog is used in local witchcraft to make a person fall gravely ill, the victim breaking out in oozing sores that follow the same pattern as the markings on the dog's coat. The concoction or *muthi* is either placed in

the person's food or sprinkled on the ground in an area he or she frequents, often outside their hut. It is evidently so potent that, in order to counteract the spell, it is necessary to consult an even more powerful sangoma.

Like the Zulu, Venda and Tswana people, and indeed all tribal groups in southern Africa, most rural Shangaan families own several hunting dogs of an ancient lineage recently recognised as *Canis africanus*. Lithe and resilient, these animals came down through Africa with their early Iron Age masters who migrated from the east and west of Africa in a series of waves, arriving in the sub-continent in the first millennium AD. Remains of the breed have been found at Iron Age sites dating back around 1600 years ago. There is also a suggestion the early Bantu peoples of the region traded their indigenous dogs with the Bushmen, as they are sometimes featured in the latter's rock art.

These dogs are generally known as *sica* and a good one is highly prized; you'll see men, usually carrying sticks, with their hunting dogs throughout rural parts of the country. Like the Zulu, Xhosa or Tswana, a Shangaan man will give his *sica* traditional medicine or muthi made from the powdered bones of a wild dog

Above Shangaan sangomas 'throw the bones' when communicating with the ancestors. The bones are called *tinhlolo*, a derivative of the word *mahlolwa*, the Shangaan word for wild dogs.

Opposite Their tracks show four digits on each foot, the short claws clearly evident. The front feet are larger than the hind.

Overleaf We watched in disbelief as a few of the wild dogs from the Manyeleti Game Reserve's so-called 'north pack' stood their ground when an elephant bull thundered towards their den. The dogs called his bluff as he stopped just metres away and sauntered off.

IN SEARCH OF THE AFRICAN WILD DOG

Above When threatened by predators a herd of impalas will scatter in all directions, running with leaps of up to three metres high and 11 metres long. The explosion of blurred shapes helps to confuse their pursuers.

to instil in them the same agility and hunting skills as *Lycaon pictus*. A hunt is generally a social occasion and the prerogative of men only and, while the women are not permitted to participate, they take a keen interest in proceedings.

There is a common belief among Shangaan men that if they should encounter a pack of wild dogs while walking in the bush, they'll be safe from attack if they lie down and 'play dead'. That's because wild dogs are supposed to eat only prey that is freshly killed and do not scavenge off carcasses – although it's not known if this has ever been put to the test! The Shangaan are part of a wider group called the Tsonga. Their language is quite melodious to the ear, so much so that it is often called the 'whistling language'. Tsonga music plays an important role in the expression of tradition and mythology. There is a popular song about wild dogs where an older child plays the character of a dog on the hunt for duiker, the part of the antelope played by several younger children, and as the song unfolds the lessons of nature are passed on.

The role of nature in nurture

The name of Dr Gus Mills is synonymous with wild dogs in the Kruger National Park and indeed far beyond its confines. He was appointed as predator biologist in the park in 1984. He began his study of dogs some five years later when, as the only endangered carnivore in Kruger, they were identified as a priority species. His research, which continued until 2006 when he left to study cheetah in the Kalahari, brought to light the most in-depth information on the dogs and contributed enormously in raising awareness of their situation and the need for urgent measures to conserve them.

In Kruger the density of wild dogs is very low in comparison with that of other large carnivores and their numbers fluctuate greatly. Food preference, habitat usage and hunting behaviour are important factors that allow them to survive and coexist with their larger and more numerous competitors. Their diet of small to medium sized antelope, particularly impala, overlaps with that of the big cats and spotted hyenas, but the diversity of mammals in the park means that the different predators can be more selective in their choice of prey. Over and above the ubiquitous impala, lionesses for instance favour wildebeest and zebra, male lions take down buffalo and giraffe, leopards go for duiker and bushbuck, warthog and primates, cheetahs also hunt duiker as well as steenbok, and spotted hyenas (like wild dogs) take kudu.

Right Although not quite as endangered as the wild dog, the long-term survival of the cheetah is also under threat. Sleek and swift, the cheetah can reach an incredible 112 kilometres an hour when chasing its prey across the open plains. The young are incorrigibly playful.

to easy theft of their kills by other predators. Here, just as elsewhere, wild dogs are particularly sensitive to kleptoparasitism (as it is called) and it can go so far as to force them to leave an area. It is thought that this caused the dramatic decline in their population in the Serengeti during the 1960s and 1970s when the number of lions and spotted hyenas was especially high.

Despite the high density of prey species on the Serengeti Plains, the tendency of the herds to migrate over long distances is not always ideal for predators as it can lead to periodic food shortages. The result is an increase in competition between the various large carnivores and the need for much bigger home ranges, all of which impact on their well-being. In spite of their own nomadic nature, wild dogs are also affected by the migratory movement of the game, particularly at denning time when a more stable prey population would be advantageous.

In contrast, the size of the Kruger National Park, which is some 20 000 square kilometres, the diversity of the terrain (encompassing 35 different landscape types), as well as the more localised movement of the antelope, seem to work in favour of the dogs by providing improved hunting opportunities, better concealment of their kills and more places to hide from other predators, particularly in the park's wooded savanna areas. It is believed that a combination of these factors has greatly contributed to the continued existence of the species in Kruger.

It is obvious that the rate of survival of their offspring affects not only individual packs, but also the state of the population as a whole. In the course of his research, Dr Mills found a fascinating correlation between rainfall and pack size in the survival of pups in their first year. He established that dry periods before and after their birth (up to the age of about nine months), as well as a larger pack size, were most beneficial to them. Hunting conditions are more favourable when it is dry and the vegetation is sparse, as the game is more visible and there is a better chance of the prey being out of condition. This makes them easier to chase and bring down and wild dogs are particularly adept at singling out any animals that are in poor health.

While their dominant prey species is impala, wild dogs will avoid areas frequented by lions, even if there are abundant herds in the vicinity, as lion predation of the dogs is high. This avoidance does not extend as much to the presence of spotted hyenas, however, as encounters between these two species are rarely fatal. The difference in habitat for both wild dogs and their prey on the Serengeti Plains, as opposed to that in the Kruger National Park, highlights the importance of environment for their survival. The open and uniform nature of the east African plains, which cover about 5 200 square kilometres, is not favourable for the dogs due

It follows that adverse hunting conditions have an effect not only on the lives of the pups back at the den, but also on the capacity of the mother to suckle them and the energy needs of the rest of the pack to hunt successfully. Good feeding would

Opposite The Kruger National Park is home to the only genetically viable population of wild dogs in South Africa. Covering some 20 000 square kilometres, it is large enough for them to roam freely and form new packs naturally.

Above Wild dogs have huge home ranges, which can cover areas of 500 to a staggering 2 000 square kilometres. Nomadic by nature and constantly on the move, the only time they stay in the same place for any length of time is when they den.

clearly also have a positive impact on the mother's general condition at the time of giving birth, as well as on the health of the new-born litter. During periods of high rainfall, grass and shrub cover is much denser, which requires additional strength to traverse and which conceals potential hazards such as holes and sharp rocks that could lead to various injuries like sprained ankles or broken legs during the hunt. Lions are more likely to successfully stalk and ambush not only their prey but the wild dogs as well – although, conversely, the dogs would be better able to hide in the thick vegetation.

After the age of six months any offspring are less likely to be influenced by recent rainfall and the size of the pack now assumes greater significance in their survival. The use of a babysitter at the den is dependent on the dogs' ability to hunt effectively without the help of the additional adult who, like the young, will also require feeding when the hunters return to the den. Leaving the pups unattended means they are vulnerable to predation. This also applies when they start moving with the pack at around three months old, as they are often hidden and watched over by a minder during the chase, their inexperience and lack of physical strength a hindrance to the group. When the youngsters are about 10 months old, they begin to run with the adults; larger packs are better equipped to provide protection during the hunt and to stand guard at the kill.

In the course of his research on the factors affecting the survival of wild dogs in the Kruger National Park, it became increasingly clear to Gus Mills that 'understanding the negative effect of preceding wet periods and the positive influence of pack size could assist in conservation planning. It could help to predict reproductive success, be used retrospectively to understand local population fluctuations and assist in the proactive management of wild dog metapopulations'.

Right When there are pups to feed, after a kill the adult dogs gulp the meat and rush back to the den to regurgitate for the youngsters. In their search for prey they can travel great distances, sometimes as much as 40 kilometres in a day.

MAHLOLWA ~ SHANGAAN

This spread A pack will occupy a den for some three months, waiting for the pups to grow, and this attracts scavengers looking for scraps. Amongst others, this hooded vulture was particularly persistent and a great irritation to these dogs.

Above As a pack of wild dogs stirs for the early morning hunt, dawn brings tension to these impala. They usually rest up during the night in the most open area available, providing any predator with the least possible cover.

Overleaf Wild dogs are adept at spotting antelope that are out of condition, making them easier to chase down. Dogs will change hunting strategy according to the terrain as well as the species of prey they are chasing.

What is the Future of Wild Dogs in the Kruger National Park?

Dr. Gus Mills comments:

Kruger is undoubtedly the most important area for wild dogs in South Africa as it contains the only viable, naturally regulated population in the country. Because of the very large areas needed to conserve wild dogs (10 000 sq km) it is very unlikely that in the future a similar sized area of wild dog habitat in South Africa can be set aside purely for large carnivore conservation. The development of the transfrontier park from the Kruger Park into Mozambique and Zimbabwe will benefit this population.

Other areas where wild dog conservation may have potential in South Africa are northern KwaZulu-Natal where corridors between a series of protected areas might be possible, and the Limpopo/Shashe Transfrontier Park from Mapungubwe into Botswana and Zimbabwe. The southern Kalahari does not appear to be a viable area for wild dogs, probably because it is too arid.

Where to See Wild Dogs in the Greater Kruger National Park

Hoedspruit Endangered Species Centre
Tel: +27(0)15-793-1633 / Fax: +27(0)15-793-1646
E-mail management@cheetahcentre.co.za
Website: www.hesc.co.za

Honeyguide Tented Safari Camps
Tel: +27(0)11-341-0282 / Fax: +27(0)11-341-0281
E-mail reservations@mix.co.za
Website: www.honeyguidecamp.com

Londolozi Game Reserve
Tel: +27(0)11-280-6655 / Fax: +27(0)11-280-6658
E-mail reservations@londolozi.co.za
Website: www.londolozi.co.za

MalaMala Game Reserve
Tel: +27(0)11-442-2267/ Fax: +27(0)11-442-2318
E-mail reservations@malamala.com
Website: www.malamala.com

Rhino Walking Safaris
Tel: +27(0)83-631-4956 / Fax: +27(0)13-735-8925
E-mail info@rws.co.za
Website: www.isibindiafrica.co.za

South African National Parks
(Kruger National Park)
Tel: +27(0)12-428-9111 / Fax: +27(0)12-343-0905
E-mail reservations@sanparks.org
Website: www.sanparks.org

Tintswalo Safari Lodge
Tel: +27(0)11-464-1070/1 / Fax: +27(0)11-464-1315
E-mail res@tintswalo.com
Website: www.tintswalo.com

References

Books

Bryant, AT. *The Zulu People* (Shuter and Shooter, Pietermaritzburg, 1949).

Davies, R, Hofmeyr, M, Dell, S, Leitner P and the Madikwe Development Task Team. *The Madikwe Development Series.* (North West Parks and Tourism Board, Mmabatho, 1997).

Estes, Richard D. *The Safari Companion* (Russell Friedman Books CC, Halfway House, 1993).

Kirby FZS, Frederick Vaughan. *In Haunts of Wild Game* (William Blockwood and Sons, Edinburgh and London, 1896).

Mills, Gus and Hess, Lex. *The Complete Book of Southern African Mammals* (Struik, Cape Town, 1997).

Mills, Michael GL and Funston, Paul J. *The Kruger Experience, Ecology and Management of Savanna Heterogeneity.* Chapter 18 (Island Press, Washington, 2003).

Napier, Cecil. *Killers and Big Game* (Howard Timmins, Cape Town, 1966).

Pringle, John. *The Conservationists and the Killers* (TV Bulpin and Books of Africa (Pty) Ltd, Cape Town, 1982).

Roodt, Veronica. *Trees and Shrubs of the Okavango Delta* (Shell Oil Botswana, Gaborone, 1998).

Scrobie, Alastair. *Animal Heaven* (Cassell and Company Ltd, London, 1953).

Skinner, JD and Smithers, RHN. *The Mammals of the Southern African Subregion* (University of Pretoria, 1990).

Stevenson-Hamilton, J. *A South African Eden: the Kruger National Park 1902–1946* (Struik, Cape Town, 1993).

Stevenson-Hamilton, J. *Wild Life in South Africa* (Cassell and Company Ltd, London, 1947).

Stokes, CS. *Sanctuary* (Maskew Miller Limited, Cape Town, 1953).

Van Dyk, Ann. *The Cheetahs of De Wildt* (The De Wildt Cheetah and Wildlife Trust, Hartbeespoort, 1991).

Wannenburgh, Alf. *The Bushmen* (Struik, Cape Town, 1979).

Wolhuter, Harry. *Memories of a Game Ranger* (The Wildlife Protection Society of South Africa, 1948).

Woodhouse, HC. *The Bushman Art of Southern Africa* (Parnell & Sons SA (Pty) Ltd, Cape Town, 1979).

Other works

Davies, Harriet T and du Toit, Johan T. 2004. Anthropogenic factors affecting wild dog *(Lycaon pictus)* re-introductions: a case study in Zimbabwe. Research paper. Oryx.

Graf, JA, Gusset, M, Reid, C, Janse van Rensberg, S, Slotow, R and Somers, MJ. 2006. Evolutionary ecology meets wildlife management: artificial group augmentation in the re-introduction of endangered African wild dogs *(Lycaon pictus)*. Research paper. Animal Conservation Journal.

Gusset, M, Maddock, AH, Gunther, GJ, Szykman, M, Slotow, R, Walters, M and Somers, MJ. 2008. Conflicting human interests over the re-introduction of endangered wild dog in South Africa. Research paper. Biodivers Conserve.

Gusset, M, Ryan, SJ, Hofmeyr, M, van Dyk, G, Davies-Mostert et al. 2007. Efforts going to the dogs? Evaluating attempts to re-introduce wild dogs in South Africa. Research paper. *Journal of Applied Ecology*, British Ecological Society.

Gusset, M, Slotow, R and Somers, MJ. 2006. Divided we fail: the importance of social integration for the re-introduction of endangered African wild dogs *(Lycaon pictus)*. Research paper. Journal of Zoology.

Havemann, Paul and Reid, Craig. Norms and standards for the management of African wild dog *(Lycaon pictus)* populations in the KwaZulu-Natal Province. Unpublished paper.

Maddock, Anthony. 1995. Wild Dog Demography in Hluhluwe Umfolozi, South Africa. Research paper.

Massicot, Paul. 2005. Animal info – African wild dog. Website.

Mills, MGL et al. 2007. Factors affecting juvenile survival in African wild dogs *(Lycaon pictus)* in Kruger National Park, South Africa. Research paper. Journal of Zoology.

Reich, Allen. 1981. The behaviour and ecology of the African wild dog *(Lycaon pictus)* in the Kruger National Park. Thesis, Yale University.

Snedegar, KV. Stars and seasons in Southern Africa. 2000. Department of Astronomy, University of Cape Town, Paper online.

Woodroffe, R, Davies-Mostert, H, Ginsberg, J, Graf, J, Leigh, K, McCreery, K, Mills, G, Pole, A, Rasmussen, G, Robbins, R, Somers, M, and Szykman, M. 2007. Rates and causes of mortality in endangered African wild dogs *(Lycaon pictus)*: lessons for management and monitoring. Research paper.

Index

Page references in italics indicate where entries appear in photographs.

A
art, wild dog *36*, *37*

B
baboons *137*
Bantich, Adrian 138, 140
Bertschinger, Henk *56*, 57, 70
bio-boundary concept 118
Bosman, Herman Charles 47
Botswana 34, 38, *40*, 41
   see also Northern Tuli Game Reserve
breeding 27–28, *27*, *28*, *29*, 31, *33*, *126*, *127*, 146, 148, *150–151*
buffalo 77
Buffalo Ridge Safari Lodge 70
Bugatsu, Patience 47–48, *49*
Bush House, The 70
Bushman folklore 38, *38–39*, *40*, 41

C
cheetahs 34, 56, 57, 145, *145*
   see also De Wildt Cheetah and Wildlife Trust
community awareness 44, 56, 78, 87, 89, 118, 120, *122*, 123, *123*
Cooper, Dave 64–65
crocodiles 34

D
Davies-Mostert, H 24, *24*, 107, 130
Dell, Steve 57, 64
De Wildt Cheetah and Wildlife Trust 9, 10, 16, 44, 49, 56–57, *56*
distemper 20, 50, 57, 70

E
eland *124–125*
elephants *42–43*, *62*, 140, *142–143*
Endangered Wildlife Trust 24, 78, 107, 123
Etali Safari Lodge 70
Ezemvelo KZN Wildlife 78, 101

F
folklore *see* traditional knowledge

G
Graaff, Carla 78, 80–81, *85*
Greater St Lucia Wetland Park
   see Isimangaliso Wetland Park
guineafowl *103*

H
Havemann, Paul 86, 100
Hluhluwe Imfolozi Park
   anti-poaching unit *94–95*
   community education 89
   contact details 101
   description and history *72–73*, 74, 77–78, *92–93*, *96*, *97*, *98*, *98–99*
   future of wild dogs 100
   KZN Wild Dog Project 78, 80–81, *84*, *85*, 100
   relocation of wild dogs 64–65, *64*, *65*, *66*, *67*
Hoedspruit Endangered Species Centre 156
Hofmeyr, Declan 52, *53*, *54–55*, 56
Hofmeyr, Markus 49, 50
Honeyguide Tented Camps 156
hunters (man) 21–22
hunting behaviour (wild dogs) 146, 148, *152–153*, *154–155*
hyena, brown 38, 41, *60*, *61*
hyena, spotted *21*, 34, 86, *91*, 146

I
impala 18, *21*, *30*, 31, *137*, *144*, *152–153*, *154–155*
Impodimo Game Lodge 70
Isimangaliso Wetland Park 86, 87

J
Jaci's Lodges 70
Jackson, C 110, *111*, 115, 118, *118*

K
Kirby, Frederick Vaughan 77, *77*
Kruger National Park 156
   description and history 22, 134, 136–137
   dogs' fear of crocodiles 34
   future of wild dogs 156
   packs in Kruger 136–137, 138, *138*, 140
   Sabie River *132–133*
   Sabi Sand Game Reserve *6–7*
   wild dogs research 145–146, *146*, *147*, 148, *148–149*, *150–151*
kudu 31, *63*, *114–115*, 115, *116*, *117*
Kwazulu-Natal
   see also Hluhluwe Imfolozi Park; Mkhuze Game Reserve; Thanda Private Game Reserve
   dispersal corridors 86–87
   future of wild dogs 100
   places to see wild dogs 101

L
leopards 34
Limpopo/Shashe Transfrontier Park 104, 106, 130, 156
Limpopo Valley
   see also Northern Tuli Game Reserve; Venetia Limpopo Nature Reserve
   future of wild dogs 130
   history 104, 106, *106*
   places to see wild dogs 131
lions
   avoidance by wild dogs 34, *58–59*, *139*, 146, 148
   Kruger National Park 145, 146
   Madikwe Game Reserve 50, *51*, 52, 57
   Phinda Resource Reserve 86
   Thanda Private Game Reserve 89
Londolozi Game Reserve 156

M
McNutt, Dr Tico 118
Madikwe Collection, The 70
Madikwe Farm House 70
Madikwe Game Reserve 70
   community awareness 44, 56
   description and history 44, 46–49, *46*, *48*
   future of wild dogs 70
   relocation from Pilanesberg 52, 64–65, *64*, *65*, *66*, *67*
   Tlou Dam *42–43*
   wild dog programme 11, 16, 35, 47, *47*, 49–50, 52, *53*, *54–55*, 56, 57
Madikwe Hills Game Lodge 70
Madikwe River Lodge 70
Madikwe Safari Lodge 70
Makanyane Safari Lodge 71
MalaMala Game Reserve 156
Manyeleti Game Reserve 138, 140, *142–143*
Mapungubwe, National Park 104, 106, 131
Mashatu Game Reserve 131
Masupe, Rex *111*, 123, *123*
Mateya Safari Lodge 71
Mills, Dr Gus 145, 146, 148, 156
Mkhuze Game Reserve 74, 86, 89, *96*, 101
Mopane Bush Lodge 131
Morukuru Lodge 70
Mosetlha Bush Camp 71
Motswiri Private Safari Lodge 70
Mpumalanga
   see also Kruger National Park
   future of wild dogs 156
   places to see wild dogs 156
Msezane, Jeremiah 87–89, *88*
Mun-ya-wana Game Reserve 74

N
Nanni, Greg 64–65
Nene, Morumo *85*
Nitani Private Game Reserve 131
Northern Tuli Game Reserve
   community attitudes 120, *122*, 123, *123*
   curiosity of dogs *19*
   den of pups *29*
   description *103*, 107
   wild dog project 106, 107, 110, *110*, *111*, *112*, *113*, *114–115*, 115, *116–117*, 118, *118*, 130

North West Province
see also Madikwe Game Reserve;
Pilanesberg National Park
future of wild dogs 70
places to see wild dogs 70–71

O
Oboletse, Kelebileone 120

P
Phinda Resource Reserve 74, 86
Phongola Private Nature Reserve 87, 100
Pilanesberg National Park
hunting behaviour *30*
reintroduction of wild dogs 57
relocation of wild dogs 52, 64–65, *64, 65, 66, 67*

R
rabies 20, 50, 57, 70
rhinoceros, white *76*, 77
Rhino Walking Safaris 156
Rhulani Safari Lodge 71
rock art 41, 134
Royal Madikwe Lodge 71

S
Sabi Sand Game Reserve *6–7*
Sasol 9, 10, 11, 16, 50, 56, 57
Selous, Frederick 21, *22*
Smidt, Emile 85
Somkhanda Game Reserve 87, 100
SANParks 131, 156
Stevenson-Hamilton, James 22, 34, 136–137, *136*

T
Tau Game Lodge 50, 71
Thakadu River Camp 70
Thanda Private Game Reserve 74, 81, 87, 89, 101
Thompson's gazelle 18, 31
Thulamela 134, 136
Tintswalo Lodge 138, 140, 156

traditional knowledge
Bushman 38, *38–39, 40*, 41
Shangaan 140–141, *141*, 145
Tswana 48–49, 120, 123
Zulu 87–89, *87, 88*
transfrontier parks
benefit for wild dogs 24
from Kruger to Mozambique/Zimbabwe 156
Limpopo/Shashe Transfrontier Park 104, 106, 130, 156
Tuli Safari Lodge 120, 131
Tuningi Safari Lodge 70

V
Van Dyk, Ann 56, *56*, 57
Van Dyk, Gus 57
Venetia Limpopo Reserve 106–107, *107, 108–109*, 123, 130, 131
Venter, Mariana 89
vulture, hooded *151*

W
Whittington-Jones, Brendan 78, 81
wild dog
appearance *17, 75, 90, 121, 140*
attitudes towards water *4–5*, 34, *80, 81, 82–83*
breeding 27–28, *27, 28, 29*, 31, *33, 126, 127*
distribution 13
endangered status *2–3*, 18
hunting behaviour *23, 25, 30*, 31, *33*, 34, 45, *58–59, 68–69, 119*
impact of man 18, 20
origins 18, *135*
perceptions 10, 16, 20
personality *19, 79*, 80
physical movement *10–11, 14–15*
prey 18, 31, *124–125, 128–129*
range 31, *147*
relationship with other predators 34, *58–59*, 86, 89, *139*, 146, 148
smell of 47, 64–65, *75*
social behaviour *8, 9, 23, 26–28, 26*, 31, *32, 34, 105, 120*
trade in 22
traditional knowledge 38, *38–39, 40*, 41, 48–49, 87–89, *87, 88*, 120, 123, 140–141, *141*, 145
Wild Dog Advisory Group (WAG) 24, 107
wild dog projects
Hluhluwe Imfolozi Game Reserve 78, 80–81, *84, 85*, 100
Kruger National Park 145–146, *146, 147*, 148, *148–149, 150–151*
Madikwe Game Reserve 16, 49–50, 52, *53, 54–55*, 56–57, 64–65, *64, 65, 66, 67*
Northern Tuli Reserve 106, 107, 110, *110, 111, 112, 113, 114–115*, 115, *116–117*, 118, *118*, 130
Venetia Limpopo Nature Reserve 106–107, *107, 108–109*, 130
wildebeest 31, *86, 128–129*
Wolhuter, Harry 22, *22*

Z
Zululand Rhino Reserve 74, 87, 100
Zwane, Zama 87, *87*, 88, 89

WHERE TO SEE WILD DOGS

Buffalo Ridge Safari Lodge 70
The Bush House 70
Etali Safari Lodge 70
Ezemvelo KZN Wildlife 101
Hluhluwe Imfolozi Park 101
Hoedspruit Endangered Species Centre 156
Honeyguide Tented Camps 156
Impodimo Game Lodge 70
Jaci's Lodges 70
Londolozi Game Reserve 156
The Madikwe Collection 70
Madikwe Farm House 70
Madikwe Game Reserve 70
Madikwe Hills Game Lodge 70
Madikwe River Lodge 70
Madikwe Safari Lodge 70
Makanyane Safari Lodge 71
MalaMala Game Reserve 156
Mapungubwe National Park 131
Mashatu Game Reserve 131
Mateya Safari Lodge 71
Mkhuze Game Reserve 101
Mopane Bush Lodge 131
Morukuru Lodge 70
Mosetlha Bush Camp 71
Motswiri Private Safari Lodge 70
Nitani Private Game Reserve 131
Rhino Walking Safaris 156
Rhulani Safari Lodge 71
Royal Madikwe Lodge 71
SANParks 131, 156
Tau Game Lodge 71
Thakadu River Camp 70
Thanda Private Game Reserve 101
Tintswalo Safari Lodge 156
Tuli Safari Lodge 131
Tuningi Safari Lodge 70
Venetia Limpopo Reserve 131